Multimedia Processing and Communications: Audio and Video

Multimedia Processing and Communications: Audio and Video

Ophelia Saunders

NY RESEARCH
P R E S S

New York

Published by NY Research Press
118-35 Queens Blvd., Suite 400,
Forest Hills, NY 11375, USA
www.nyresearchpress.com

Multimedia Processing and Communications: Audio and Video
Ophelia Saunders

International Standard Book Number: 978-1-63238-699-1 (Hardback)

Cataloging-in-Publication Data

Multimedia processing and communications : audio and video / Ophelia Saunders.
    p. cm.
Includes bibliographical references and index.
ISBN 978-1-63238-699-1
1. Multimedia communications. 2. Telecommunication. 3. Multimedia systems.
4. Sound--Recording and reproducing--Digital techniques.
5. Digital video. I. Saunders, Ophelia.
TK5105.15 .M85 2019
621.382--dc23

Contents

Permissions

Index

Preface

Multimedia refers to the content that uses text, audio, video, image, graphics and animation to create content. The purpose of multimedia is to enhance the users' experience and generate interactivity by combining diverse forms of media content. Multimedia content is stored in multimedia devices. It can be in the form of presentations, games and simulations. Online multimedia is object-oriented and data-driven, which allows user innovation and personalization. Haptic technology further upgrades multimedia experience by allowing virtual objects to be felt. Various multimedia tools are used in advertising, computer-based training, developing special effects in movies and video games, digital media, computer simulations, etc. This textbook provides comprehensive insights in the domain of multimedia technology. Most of the topics introduced in this book cover new techniques and applications of this technology. It will provide comprehensive knowledge to the readers.

A foreword of all chapters of the book is provided below:

Chapter 1 - An audio signal typically represents a sound wave as an electrical voltage or as a binary number, depending on whether it is an analog or digital signal. The topics of significance in the understanding of audio signals, such as digital audio, noise gate, equalization and alignment level, have been covered in extensive detail in this chapter; **Chapter 2** - Audio signal processing is the alteration of audio signals by using an effects unit or an audio effect. This is an introductory chapter, which will introduce briefly all the significant aspects of audio processing. It includes a number of significant topics such as dynamic range compression, audio mastering, audio mixing, audio post production, sound effect, etc.; **Chapter 3** - Video signals can be both analog and digital. This chapter discusses in detail the different types of video signals, such as component video and composite video, as well as the aspects of video quality, video scaler, video file, coding format, etc.; **Chapter 4** - Video processing is a form of signal processing, which implements video filters for getting video streams or video files, for use in TV sets, DVDs, VCRs, video players, etc. This chapter has been carefully written to provide an easy understanding of the varied aspects of video processing such as scan conversion, video capture, deinterlacing, color grading, motion interpolation, etc.; **Chapter 5** - Video editing refers to the arrangement and manipulation of video shots. It is significant in films and television shows, video essays and video advertisements. The various technological innovations in video editing, such as continuity editing, linear and non-linear editing systems have been covered in great detail in this chapter.

At the end, I would like to thank all the people associated with this book devoting their precious time and providing their valuable contributions to this book. I would also like to express my gratitude to my fellow colleagues who encouraged me throughout the process.

Ophelia Saunders

Audio Signals

An audio signal typically represents a sound wave as an electrical voltage or as a binary number, depending on whether it is an analog or digital signal. The topics of significance in the understanding of audio signals, such as digital audio, noise gate, equalization and alignment level, have been covered in extensive detail in this chapter.

Audio signals are generally referred to as signals that are audible to humans. Audio signals usually come from a sound source which vibrates in the audible frequency range. The vibrations push the air to form pressure waves that travels at about 340 meters per second. Our inner ears can receive these pressure signals and send them to our brain for further recognition.

There are numerous ways to classify audio signals. If we consider the source of audio signals, we can classify them into two categories:

- Sounds from animals: Such as human voices, dog's barking, cat's mewing, frog's croaking. (In particular, Bioacoustics is a cross-disciplinary science, which investigates sound production and reception in animals.)

- Sounds from non-animals: Sounds from car engines, thunder, door slamming, music instruments, etc.

We can divide each short segment (also known as frame, with a length of about 20 ms) of human's voices into two types:

- Voiced sound: These are produced by the vibration of vocal cords. Since they are produced by the regular vibration of the vocal cords, you can observe the fundamental periods in a frame. Moreover, due to the existence of the fundamental period, you can also perceive a stable pitch.

- Unvoiced sound: These are not produced by the vibration of vocal cords. Instead, they are produced by the rapid flow of air through the mouse, the nose, or the teeth. Since these sounds are produced by the noise-like rapid air flow, we can not observed the fundamenta period and no stable pitch can be perceived.

It is very easy to distinguish between these two types of sound. When you pronunce an utterance, just put your hand on your throat to see if you feel the vibration of your vocal cords. If yes, it is voiced; otherwise it is unvoiced. You can also observe the waveforms to see if you can identify the fundamental periods. If yes, it is voiced; otherwise, it is unoviced.

Signal Flow

Signal flow is the path an audio signal will take from source to the speaker or recording device. Signal flow may be short and simple as in a home audio system or long and convoluted in a recording studio and larger sound reinforcement system as the signal may pass through many sections of a large console, external audio equipment, and even different rooms.

Parameters

Audio signals may be characterized by parameters such as their bandwidth, nominal level, power level in decibels (dB), and voltage level. The relation between power and voltage is determined by the impedance of the signal path, which may be single-ended or balanced.

Audio signals have somewhat standardized levels depending on application. Outputs of professional mixing consoles are most commonly at line level. Consumer audio equipment will also output at a lower line level. Microphones generally output at an even lower level, commonly referred to a *mic level*.

Digital Equivalent

As much of the older analog audio equipment has been emulated in digital form, usually through the development of audio plug-ins for digital audio workstation (DAW) software, the path of digital information through the DAW (i.e. from an audio track through a plug-in and out a hardware output) is also called an *audio signal* or *signal flow*.

A digital audio signal being sent through wire can use several formats including optical (ADAT, TDIF), coaxial (S/PDIF), XLR (AES/EBU), and Ethernet.

Digital Recording

Digital sound recording, method of preserving sound in which audio signals are transformed into a series of pulses that correspond to patterns of binary digits (i.e., 0's and 1's) and are recorded as such on the surface of a magnetic tape or optical disc. A digital system samples a sound's wave form, or value, several thousand times a second and assigns numerical values in the form of binary digits to its amplitude at any given instant. A typical digital recording system is equipped with an analog-to-digital converter that transforms two channels of continuous audio signals into digital information, which is then recorded by a high-speed tape or disc machine. The system uses a digital-to-analog converter that reads the encoded information from the recording medium and changes it back into audio signals that can be used by the amplifier of a conventional stereo sound system.

Digital recording provides higher-fidelity sound reproduction than do ordinary recording methods, largely because audio signals converted into simple pulse patterns are virtually immune to the residual noise and distortion that are characteristic of analog communication channels and sound recording media. In addition, many digital recording systems are designed to detect and eliminate interfering signals. In the 1980s digital compact disc recordings became available that were played by using a laser beam to optically scan digital information encoded on the disc's surface. In the late 1980s digital audio tape (DAT) recorders using magnetic tape cassettes became available for audio reproduction and recording. The DAT recorder converts audio signals into digital data on a magnetic tape by means of a microprocessor and converts the data back into analog audio signals that can be used by the amplifier of a conventional stereo sound system. The early 1990s saw the introduction of digital compact cassette (DCC) recorders, which were similar to DAT recorders but could play the older analog tape cassettes in addition to similarly shaped digital cassettes.

Process

Recording

1. The analog signal is transmitted from the input device to an analog-to-digital converter (ADC).

2. The ADC converts this signal by repeatedly measuring the momentary level of the analog (audio) wave and then assigning a binary number with a given quantity of bits (word length) to each measuring point.

3. The frequency at which the ADC measures the level of the analog wave is called the sample rate or sampling rate.

4. A digital audio sample with a given word length represents the audio level at one moment.

5. The longer the word length the more precise the representation of the original audio wave level.

6. The higher the sampling rate the higher the upper audio frequency of the digitized audio signal.

7. The ADC outputs a sequence of digital audio samples that make up a continuous stream of 0s and 1s.

8. These binary numbers are stored on recording media such as a hard drive, optical drive or in solid state memory.

Playback

1. The sequence of numbers is transmitted from storage into a digital-to-analog converter (DAC), which converts the numbers back to an analog signal by

sticking together the level information stored in each digital sample, thus rebuilding the original analog wave form.

2. This signal is amplified and transmitted to the loudspeakers or video screen.

Recording of Bits

Even after getting the signal converted to bits, it is still difficult to record; the hardest part is finding a scheme that can record the bits fast enough to keep up with the signal. For example, to record two channels of audio at 44.1 kHz sample rate with a 16 bit word size, the recording software has to handle 1,411,200 bits per second.

Techniques to Record to Commercial Media

For digital cassettes, the read/write head moves as well as the tape in order to maintain a high enough speed to keep the bits at a manageable size.

For optical disc recording technologies such as CDs or DVDs, a laser is used to burn microscopic holes into the dye layer of the medium. A weaker laser is used to read these signals. This works because the metallic substrate of the disc is reflective, and the unburned dye prevents reflection while the holes in the dye permit it, allowing digital data to be represented.

Concerns with Digital Audio Recording

Word Size

The number of bits used to represent a sampled audio wave (the *word size*) directly affects the resulting noise in a recording after intentionally added dither, or the distortion of an undithered signal.

The number of possible voltage levels at the output is simply the number of levels that may be represented by the largest possible digital number (the number 2 raised to the power of the number of bits in each sample). There are no "in between" values allowed. If there are more bits in each sample the waveform is more accurately traced, because each additional bit doubles the number of possible values. The distortion is roughly the percentage that the least significant bit represents out of the average value. Distortion (as a percentage) in digital systems increases as signal levels decrease, which is the opposite of the behavior of analog systems.

Sample Rate

The sample rate is just as important a consideration as the word size. If the sample rate is too low, the sampled signal cannot be reconstructed to the original sound signal.

To overcome aliasing, the sound signal (or other signal) must be sampled at a rate at least twice that of the highest frequency component in the signal. This is known as the Nyquist-Shannon sampling theorem.

For recording music-quality audio the following PCM sampling rates are the most common: 44.1, 48, 88.2, 96, 176.4, and 192 kHz.

When making a recording, experienced audio recording and mastering engineers will normally do a master recording at a higher sampling rate (i.e. 88.2, 96, 176.4 or 192 kHz) and then do any editing or mixing at that same higher frequency. High resolution PCM recordings have been released on DVD-Audio (also known as DVD-A), DAD (Digital Audio Disc—which utilizes the stereo PCM audio tracks of a regular DVD), DualDisc (utilizing the DVD-Audio layer), or Blu-ray (Profile 3.0 is the Blu-ray audio standard, although as of mid-2009 it is unclear whether this will ever really be used as an audio-only format). In addition it is nowadays also possible and common to release a high resolution recording directly as either an uncompressed WAV or lossless compressed FLAC file (usually at 24 bits) without down-converting it.

However, if a CD (the CD Red Book standard is 44.1 kHz 16 bit) is to be made from a recording, then doing the initial recording using a sampling rate of 44.1 kHz is obviously one approach. Another approach that is usually preferred is to use a higher sample rate and then downsample to the final format's sample rate. This is usually done as part of the mastering process. One advantage to the latter approach is that way a high resolution recording can be released, as well as a CD and/or lossy compressed file such as mp3—all from the same master recording.

Beginning in the 1980s, music that was recorded, mixed and mastered digitally was often labelled using the SPARS code to describe which processes were analog and which were digital.

Digital Audio

Digital audio is a technology that is used to record, store, manipulate, generate and reproduce sound using audio signals that have been encoded in digital form.

It also refers to the sequence of discreet samples that are taken from an analog audio waveform. Instead of a continuous sinusoidal wave, digital audio is composed of discreet points which represent the amplitude of the waveform approximately.

The more samples taken, the better the representation, and hence impacts the quality of the digital audio. Most modern multimedia devices can only process digital audio, and in the case of cellphones requiring analog audio input, they still convert it to digital before transmission.

To create a digital audio from an analog audio source, tens of thousands of samples are taken per second to ensure the replication of the waveform, with each sample representing the intensity of the waveform in that instant.

The samples are stored in binary form same as any digital data, regardless of the type. The samples which are merged into a single data file must be formatted correctly in order for it to be played on a digital player with the most common digital audio format being MP3.

Apart from the sampling frequency, another parameter in digital encoding is the number of bits used when taking samples. The common sampling parameter used is 16 bit samples taken over a spectrum of 44.1 thousand cycles per second or 44.1 Kilo Hertz (kHz). CD quality digital audio therefore requires 1.4 million bits of data per second.

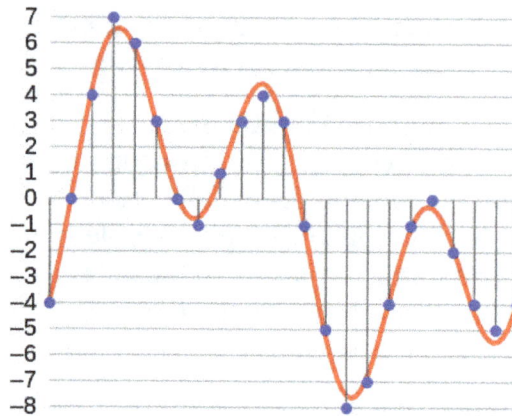

A sound wave, in red, represented digitally, in blue (after sampling and 4-bit quantization).

Digital audio technologies are used in the recording, manipulation, mass-production, and distribution of sound, including recordings of songs, instrumental pieces, podcasts, sound effects, and other sounds. Modern online music distribution depends on digital recording and data compression. The availability of music as data files, rather than as physical objects, has significantly reduced the costs of distribution. Before digital audio, the music industry distributed and sold music by selling physical copies in the form of records and cassette tapes. With digital-audio and online distribution systems such as iTunes, companies sell digital sound files to consumers, which the consumer receives over the Internet.

An analog audio system converts physical waveforms of sound into electrical representations of those waveforms by use of a transducer, such as a microphone. The sounds are then stored on an analog medium such as magnetic tape, or transmitted through an analog medium such as a telephone line or radio. The process is reversed for reproduction: the electrical audio signal is amplified and then converted back into physical waveforms via a loudspeaker. Analog audio retains its fundamental wave-like characteristics throughout its storage, transformation, duplication, and amplification.

Analog audio signals are susceptible to noise and distortion, due to the innate characteristics of electronic circuits and associated devices. Disturbances in a digital system do not result in error unless the disturbance is so large as to result in a symbol being misinterpreted as another symbol or disturb the sequence of symbols. It is therefore generally possible to have an entirely error-free digital audio system in which no noise or distortion is introduced between conversion to digital format, and conversion back to analog.

A digital audio signal may optionally be encoded for correction of any errors that might occur in the storage or transmission of the signal. This technique, known as channel coding, is essential for broadcast or recorded digital systems to maintain bit accuracy. Eight-to-fourteen modulation is a channel code used in the audio compact disc (CD).

Conversion Process

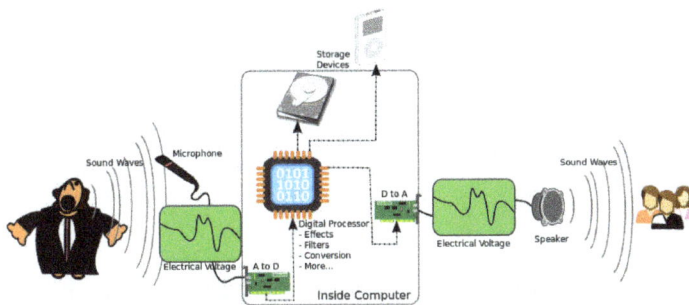

The lifecycle of sound from its source, through an ADC, digital processing,
\ a DAC, and finally as sound again.

A digital audio system starts with an ADC that converts an analog signal to a digital signal. The ADC runs at a specified sampling rate and converts at a known bit resolution. CD audio, for example, has a sampling rate of 44.1 kHz (44,100 samples per second), and has 16-bit resolution for each stereo channel. Analog signals that have not already been bandlimited must be passed through an anti-aliasing filter before conversion, to prevent the aliasing distortion that is caused by audio signals with frequencies higher than the Nyquist frequency (half the sampling rate).

A digital audio signal may be stored or transmitted. Digital audio can be stored on a CD, a digital audio player, a hard drive, a USB flash drive, or any other digital data storage device. The digital signal may be altered through digital signal processing, where it may be filtered or have effects applied. Sample-rate conversion including up-sampling and downsampling may be used to conform signals that have been encoded with a different sampling rate to a common sampling rate prior to processing. Audio data compression techniques, such as MP3, Advanced Audio Coding, Ogg Vorbis, or FLAC, are commonly employed to reduce the file size. Digital audio can be carried over digital audio interfaces such as AES3 or MADI. Digital audio can be carried over a network using audio over Ethernet, audio over IP or other streaming media standards and systems.

For playback, digital audio must be converted back to an analog signal with a DAC which may use oversampling.

Audio Coding Format

Hundreds of file formats exist for recording and playing digital sound and music files. While many of these file formats are software dependant — for example a Creative Labs Music File is a.cmf — there are several well-known and widely supported file formats. While different operating systems have different popular music file formats, we'll mainly focus on those that are most commonly used on Windows-based PCs.

Many different digital audio formats and different software are used to create, store and manipulate these files,

Should you not have the correct device for playing a particular file, you can also look for software conversion tools that will convert one file type to another. Because some audio files are open standards and some are proprietary, chances are we'll be seeing a wide variety of digital audio formats for some time to come.

File Format and Codec

An audio file format and audio codec (compressor/decompressor) are two very different things. Audio codecs are the libraries that are executed in multimedia players. The audio codec is actually a computer program that compresses or decompresses digital audio data according to the audio file format specifications. For example, the WAV audio file format is usually coded in the OCM format, as are the popular Macintosh AIFF audio files.

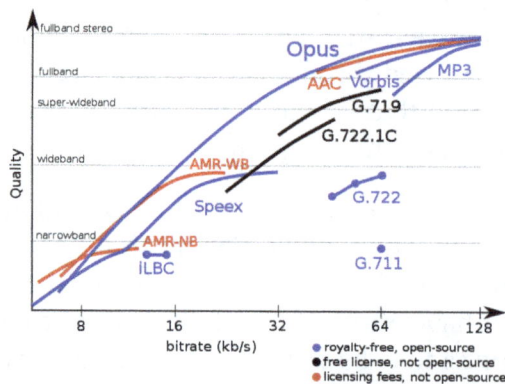

Comparison of coding efficiency between popular audio formats

An audio coding format (or sometimes audio compression format) is a content representation format for storage or transmission of digital audio (such as in digital television, digital radio and in audio and video files). Examples of audio coding formats include MP3, AAC, Vorbis, FLAC, and Opus. A specific software or hardware implementation capable of audio compression and decompression to/from a specific audio coding format is called an audio codec; an example of an audio codec is LAME, which is one of several different codecs which implements encoding and decoding audio in the MP3 audio coding format in software.

Some audio coding formats are documented by a detailed technical specification document known as an audio coding specification. Some such specifications are written and approved by standardization organizations as technical standards, and are thus known as an audio coding standard. The term "standard" is also sometimes used for de facto standards as well as formal standards.

Audio content encoded in a particular audio coding format is normally encapsulated within a container format. As such, the user normally doesn't have a raw AAC file, but instead has a.m4a audio file, which is a MPEG-4 Part 14 container containing AAC-encoded audio. The container also contains metadata such as title and other tags, and perhaps an index for fast seeking. A notable exception is MP3 files, which are raw audio coding without a container format. De facto standards for adding metadata tags such as title and artist to MP3s, such as ID3, are hacks which work by appending the tags to the MP3, and then relying on the MP3 player to recognize the chunk as malformed audio coding and therefore skip it. In video files with audio, the encoded audio content is bundled with video (in a video coding format) inside a multimedia container format.

An audio coding format does not dictate all algorithms used by a codec implementing the format. An important part of how lossy audio compression works is by removing data in ways humans can't hear, according to a psychoacoustic model; the implementer of an encoder has some freedom of choice in which data to remove (according to their psychoacoustic model).

Lossless, Lossy, and Uncompressed Audio Coding Formats

A lossless audio coding format reduces the total data needed to represent a sound but can be de-coded to its original, uncompressed form. A lossy audio coding format additionally reduces the bit resolution of the sound on top of compression, which results in far less data at the cost of irretrievably lost information.

Consumer audio is most often compressed using lossy audio codecs as the smaller size is far more convenient for distribution. Lossless audio coding formats such as FLAC and Apple Lossless are sometimes available, though at the cost of larger files.

Uncompressed audio formats, such as pulse-code modulation (.wav), are also sometimes used.

Noise Gate

A noise gate allows a signal above a certain selected threshold to pass through, known as an open gate. If the input signal falls below the threshold, the signal gets cut off and consequently no sound is heard, known as a closed gate.

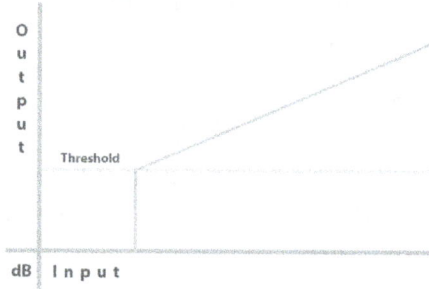

Noise gates are usually used in the following situations:-

1. Reduce background noise in a recording.

2. Help eliminate leakage between instruments such as drums.

There is generally no benefit in gating sampled drums and instruments, as there is usually no noise present. Many digital reverb units allow the reverb to be gated.

Common controls on a noise gate include threshold, attack, release, hold and range.

Threshold- The threshold sets the level at which the gate opens and closes.

Attack- The attack determines how quickly the gate opens. Measured in milliseconds

Release- The release determines how quickly the gate closes. Measured in milliseconds

Hold- The hold control determines how long the gate is kept open for after the signal falls below the threshold level, and helps to ensure that a decayed proportion of a sound is not cut off.

Range- The range control turns a signal down that is under the threshold instead of muting it completely.

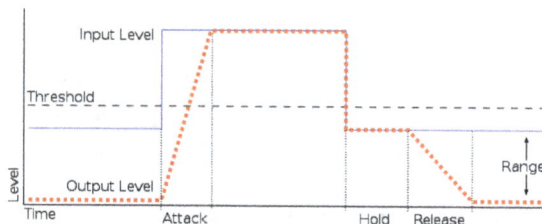

The attack, hold, and release functions of a noise gate.

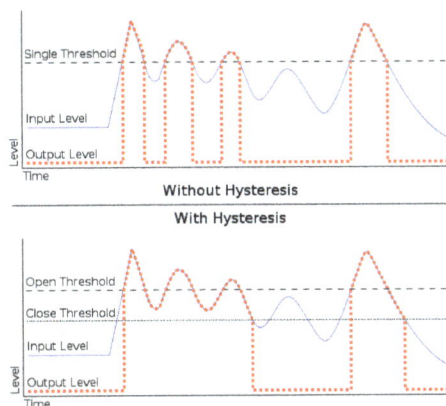

A noise gate in OBS without hysteresis can open and close undesirably with a fluctuating signal (top). With hysteresis, the noise gate does not 'chatter'.

Advanced gates have an external sidechain. This is an additional input that allows the gate to be triggered by another audio signal. A variation of a sidechained noise gate used in electronic music production is a trancegate or just simply gate, where the noise gate is not controlled by audio signal but a preprogrammed pattern, resulting in a precisely controlled chopping of a sustained sound.

Noise gates often implement hysteresis, that is, they have two thresholds: one to open the gate and another, set a few dB below, to close the gate. This means that once a signal has dropped below the close threshold, it has to rise to the open threshold for the gate to open, so that a signal that crosses over the close threshold regularly does not open the gate and cause chattering. A longer hold time also helps to avoid chattering, as described above.

Roles

A stompbox-format gate designed for use with electric guitar.

The basic function of a noise gate is to eliminate sounds below a given threshold. Noise gates are commonly used in the recording studio and sound reinforcement. Rock musicians may also use small portable "stompbox" units to control unwanted noise from their guitar amplification systems. Band-limited noise gates are also used to eliminate background noise from audio recordings by eliminating frequency bands that contain only static.

Audio Noise Reduction

In audio post-processing, noise gating reduces steady noise sources such as rumble from LP records, hiss from audio tape, static from a radio or amplifier, and hum from a power system, without greatly affecting the source sound. An audio signal such as music or speech is broken up into many frequency bands by a collection of overlapping band-pass filters, and if the signal amplitude in any one band is lower than a preset threshold, then that band is eliminated from the final sound. This greatly reduces perceptible background noise because only the frequency components of the noise that are within the gated passbands survive.

The technique was implemented in real-time electronics in some audiophile record players as early as the 1980s, and is now commonly used in audio production post-processing, where software to Fourier transform the audio signal can yield a very detailed spectrum of the background noise. Common digital audio editing software packages such as CoolEdit and Audacity include easy-to-use digital noise gating code: the user selects a segment of audio that contains only static, and the amplitude levels in each frequency band are used to determine the threshold levels to be applied across the signal as a whole.

Noise gating works well when the static is steady and either narrowly confined in frequency (e.g. hum from AC power) or well below the main signal level (15 dB minimum is desirable). In cases where the signal merges with the background static (for example, the brushed drum sounds in the Sun King track on the Beatles album Abbey Road) or is weak compared to the noise (as in very faint tape recordings), the noise gating can add artifacts that are more distracting than the original static.

In the context of a multi-microphone recording session, noise gating is employed to reduce the leakage of sound into a microphone from sources other than the one the microphone was intended for. One example involves the mic-ing up of a drumkit. In most multi-mic drum recordings, one microphone will be used to capture the snare drum sound and another to capture the kick drum sound. The snare microphone will output a signal composed of a high level snare signal and a lower level kick drum signal (due to the further distance of the kick drum from the snare microphone). If the threshold level of the noise gate is set correctly, a snare drum signal can be isolated. To fully isolate the snare drum signal, the release rate has to be quite fast, which can cause the tail end of the snare sound to be 'chopped off'. This can usually be remedied by the inclusion of one or more overhead microphones, which can act as a general 'audio glue' for all the other gated sources.

Noise gates are useful when editing live recordings in order to remove background noise between passages or pieces of dialogue. However, care must be taken in setting the gates so they do not trigger due to spurious noise such as when people move chairs.

For vocal applications on stage, an optical microphone switch may be used. An infra-red sensor senses if somebody is in front of the microphone and switches on the microphone.

Recording Usages

A good example of time-controlled noise gating is the well-known "gated reverb" effect heard on the drums on the Phil Collins hit single "In the Air Tonight", created by engineer-producer Hugh Padgham, in which the powerful reverberation added to the drums is cut off by the noise gate after a few milliseconds, rather than being allowed to decay naturally. This can also be achieved by: sending the 'dry' snare signal to the reverb (or other process) unit, inserting a noise gate on the path of the reverb signal and connecting the snare sound to the side chain of the gate unit. With the gate unit set to 'external sidechain' (or 'external key'), the gate will respond to the snare signal level and 'cut off' when that has decayed below the threshold, not the reverberated sound.

It is a common production trick to use spurious combinations of side chain inputs to control longer, more sustained sounds. For example, a hi-hat signal can be used to control a sustained synthesized sound to produce a rhythmic melodic (or harmonic) signal which is perfectly in time with the hi-hat signal. A good example of this use of the device can be found on the Godley & Creme concept album *Consequences*. The album's story required the creation of a number of special sound effects that would convey the impression of natural disasters.

For the "Fire" sequence, Godley and Creme used a noise gate, triggered by the sound of multitracked voices, that created the 'voice' of a raging bushfire. During the recording of this segment, each time the voice signal began, it triggered the noise gate to open up another channel, which carried a pre-recorded loop of a crackling sound (created by overdubbing the sound of Bubble Wrap being popped in front of a microphone). The combined voices and crackling created an eerie and quite convincing 'talking fire' effect.

Multi-latch Gating

The invention of a technique, called multi-latch gating by Jay Hodgson, common in classical music recordings for years, is often credited to producer Tony Visconti, whose use on David Bowie's "Heroes" may have been the first in rock. Visconti recorded Bowie's vocals in a large space using three microphones placed 9 inches (23 cm), 20 feet (6.1 m), and 50 feet (15.2 m) away, respectively. A different gate was applied to each microphone so that the farther microphone was triggered only when Bowie reached the appropriate volume, and each microphone was muted as the next one was triggered.

Trance Gating

Envelope following (also called trance gating because of its prevalence in trance) is the use of a gate on a track additional to the one it attenuates, so called because the latter's amplitude profile will then match or closely follow that of the first. Envelope following may be used to create syncopated rhythms or to tighten sloppy performances. For example, a synth pad may play whole notes while keyed to a guitar or percussion part. Examples include DJ Nexus's "Journey Into Trance" (1:11), Le Chic's "Everybody Dance", and Diana Ross's "Upside Down".

Sound Reinforcement

Noise gates play an important role in drum mic'ing in heavy metal shows. The drum and cymbal mic channels will typically have noise gates so that the mics will only be turned on when the specific drum or cymbal is being played. This dramatically reduces "bleeding" between the drum mics. This is used to reduce audio feedback triggered by other highly amplified sounds onstage.

Noise Reduction in Audio

Noise Reduction can reduce constant background sounds such as hum, whistle, whine, buzz, and "hiss", such as tape hiss, fan noise or FM/webcast carrier noise. It is not suitable for individual clicks and pops, or irregular background noise such as from traffic or an audience.

To use Noise Reduction, you need a region in the waveform that contains only the noise you want to reduce.

Be aware that it may be impossible to get a satisfactory removal when the noise is very loud, when the noise is variable, when the music or speech is not much louder than the noise or when the noise frequencies are very similar to those of the music or speech.

If your problem is mains hum or a high-pitched whistle, the use of a Notch Filter may help, which should be carried out before applying Noise Reduction.

Audio Equalization

The most basic type of equalization familiar to most people is the treble/bass control on home audio equipment. The treble control adjusts high frequencies, the bass control adjusts low frequencies. This is adequate for very rudimentary adjustments — it only provides two controls for the entire frequency spectrum, so each control adjusts a fairly wide range of frequencies.

Advanced equalization systems provide a fine level of frequency control. The key is to be able to adjust a narrower range of frequencies without affecting neighbouring frequencies.

Equalization is most commonly used to correct signals which sound unnatural. For example, if a sound was recorded in a room which accentuates high frequencies, an equalizer can reduce those frequencies to a more normal level. Equalization can also be used for applications such as making sounds more intelligible and reducing feedback.

Filter Types

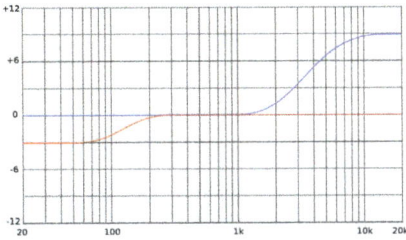

Two first-order shelving filters: a -3 dB bass cut (red), and a +9 dB treble boost (blue)

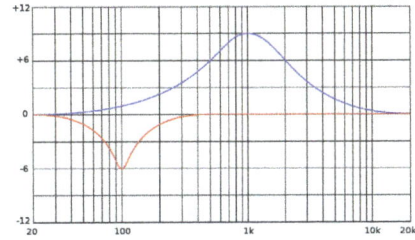

Second-order linear filter functions. Blue: a 9 dB boost at 1 kHz. Red: a 6 dB cut at 100 Hz having a higher Q (sharper bandwidth)

Although the range of equalization functions is governed by the theory of linear filters, the adjustment of those functions and the flexibility with which they can be adjusted varies according to the topology of the circuitry and controls presented to the user. Shelving controls are usually simple first-order filter functions which alter the relative gains between frequencies much higher and much lower than the cutoff frequencies. A *low shelf*, such as the bass control on most hi-fi equipment, is adjusted to affect the gain of lower frequencies while having no effect well above its cutoff frequency. A *high shelf*, such as a treble control, adjusts the gain of higher frequencies only. These are broad adjustments designed more to increase the listener's satisfaction than to provide actual equalization in the strict sense of the term.

A parametric equalizer, on the other hand, has one or more sections each of which implements a second-order filter function. This involves three adjustments: selection of the center frequency (in Hz), adjustment of the Q which determines the sharpness of the bandwidth, and the level or gain control which determines how much those frequencies are boosted or cut relative to frequencies much above or below the center frequency selected. In a *semi-parametric* equalizer there is no control for the bandwidth (it is preset by the designer) or is only selected between two presets using a switch. In a *quasi-parametric* equalizer, the bandwidth is depending on the gain level. With rising gain, the bandwidth gets wider.

A graphic equalizer also implements second-order filter functions in a more user-friendly manner, but with somewhat less flexibility. This equipment is based on a bank of filters covering the audio spectrum in up to 30 frequency bands. Each second-order filter has a fixed center frequency and Q, but an adjustable level. The user can raise or lower each slider in order to visually approximate a "graph" of the intended frequency response.

Since "equalization" in the context of audio reproduction isn't used strictly to compensate for the deficiency of equipment and transmission channels, the use of high and low pass filters may be mentioned. A high-pass filter modifies a signal only by eliminating lower frequencies. Thus a low-cut or rumble filter is used to remove infrasonic energy from a program which may consume undue amplifier power and cause excessive excursions in (or even damage to) speakers. A low-pass filter only modifies the audio signal by removing high frequencies. Thus a high-cut or hiss filter may be used to remove annoying white noise at the expense of the crispness of the program material.

A first-order low or high pass filter has a standard response curve which reduces the unwanted frequencies well above or below the cutoff frequency with a slope of 6 dB per octave. A second-order filter will reduce those frequencies with a slope of 12 dB per octave and moreover may be designed with a higher Q or finite zeros in order to effect an even steeper response around the cutoff frequency. For instance, a second-order *low-pass notch* filter section only reduces (rather than eliminates) very high frequencies, but has a steep response falling to zero at a specific frequency (the so-called *notch frequency*). Such a filter might be ideal, for instance, in completely removing the 19 kHz FM stereo subcarrier pilot signal while helping to cut even higher frequency subcarrier components remaining from the stereo demultiplexer.

In addition to adjusting the relative amplitude of frequency bands, an audio equalizer may alter the relative phases of those frequencies. While the human ear is not as sensitive to the phase of audio frequencies (involving delays of less than 1/30 second), music professionals may favor certain equalizers because of how they affect the timbre of the musical content by way of audible phase artifacts.

High-pass and Low-pass Filters

A high-pass filter is a filter, an electronic circuit or device, that passes higher frequencies well but attenuates (cuts or decreases) lower frequency components. A low-pass filter passes low-frequency components of signals while attenuating higher frequencies. Some audiophiles use a low-pass filter in the signal chain before their subwoofer speaker enclosure, to ensure that only deep bass frequencies reach the subwoofer. In audio applications these are frequently termed "low cut" and "high cut" respectively, to emphasize their effect on the original signal. For instance, sometimes audio equipment will include a switch labeled "high cut" or described as a "hiss filter" (hiss being high-frequency noise). In the phonograph era, many stereos would include a switch to introduce a high-pass (low cut) filter, often called a "rumble filter", to eliminate infra-sonic frequencies.

Shelving Filter

While high and low pass filters are useful for removing unwanted signal above or below a set frequency, shelving filters can be used to reduce or increase signals above or below a set frequency. Shelving filters are used as common tone controls (bass and treble) found in consumer audio equipment such as home stereos, and on guitar amplifiers and bass amplifiers. These implement a first order response and provide an adjustable boost or cut to frequencies above or lower than a certain point.

A *high shelf* or "treble control" will have a frequency response $|H(f)|$ whose square is given by:

$$|H(f)|^2 = \frac{1+(f/f_z)^2}{1+(f/f_p)^2}$$

where f_p and f_z are called the pole and zero frequencies, respectively. Turning down the treble control increases f_z and decreases f_p so that frequences higher than f_p are atten-uated. Turning up the treble control increases f_p and decreases f_z so that frequencies higher than f_z are boosted. Setting the treble control at the center sets $f_z = f_p$ so that $|H(f)|^2=1$ and the circuit has no effect. At most, the slope of the filter response in the transition region will be 6 dB per octave (thus a doubling of signal voltage and a conse-quent quadrupling of signal power for every doubling of frequency).

Similarly the response of a *low shelf* (or "low shelving or "bass control") can be repre-sented as

$$|H(f)|^2 = (f_z/f_p)^2 \frac{1+(f/f_z)^2}{1+(f/f_p)^2} .$$

In this case the inclusion of the leading factor simply indicates that the response at fre-quencies much higher than f_z or f_p is unity and that only bass frequencies are affected.

Note that a high shelving control in which f_z is set to infinity or a low shelve response in which f_z is set to zero implements a first order low-pass or high-pass filter respec-tively. However usual tone controls have a more limited range, since the purpose isn't to eliminate any frequencies but only to achieve a greater balance when, for instance, the treble is lacking and the sound is not crisp. Since the range of possible responses from shelving filters is so limited, some audio engineers considered shelving controls inadequate for equalization tasks.

On some bass amps and DI boxes, the units provide both low and high shelving con-trols and additional equalization controls.

Graphic Equalizer

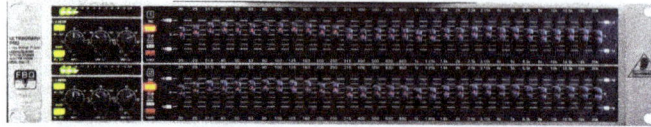

In the *graphic equalizer*, the input signal is sent to a bank of filters. Each filter passes the portion of the signal present in its own frequency range or *band*. The amplitude passed by each filter is adjusted using a slide control to boost or cut frequency components passed by that filter. The vertical position of each slider thus indicates the gain applied at that frequency band, so that the knobs resemble a *graph* of the equalizer's response plotted versus frequency.

UREI graphic and parametric EQs

The number of frequency channels (and therefore each one's bandwidth) affects the cost of production and may be matched to the requirements of the intended application. A car audio equalizer might have one set of controls applying the same gain to both stereo channels for convenience, with a total of five to ten frequency bands. On the other hand, an equalizer for professional live sound reinforcement typically has some 25 to 31 bands, for more precise control of feedback problems and equalization of room modes. Such an equalizer is called a 1/3-octave equalizer (spoken informally as "*third-octave* EQ") because the center frequency of its filters are spaced one third of an octave apart, three filters to an octave. Equalizers with half as many filters per octave are common where less precise control is required—this design is called a 2/3-octave equalizer.

Parametric Equalizer

Parametric equalizers are multi-band variable equalizers which allow users to control the three primary parameters: amplitude, center frequency and bandwidth. The amplitude of each band can be controlled, and the center frequency can be shifted, and bandwidth (which is inversely related to "Q") can be widened or narrowed. Parametric equalizers are capable of making much more precise adjustments to sound than other equalizers, and are commonly used in sound recording and live sound reinforcement. Parametric equalizers are also sold as standalone outboard gear units.

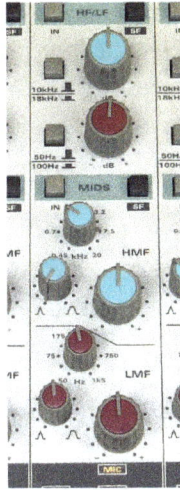

The equaliser section from the Audient ASP8024 Mixing console. The upper section has high and low shelving EQ, the lower section has fully parametric EQ.

A variant of the parametric equalizer is the semi-parametric equalizer, also known as a sweepable filter. It allows users to control the amplitude and frequency, but uses a preset bandwidth of the center frequency. In some cases, semi-parametric equalizers allow the user to select between a wide and a narrow preset bandwidth.

Filter Functions

The responses of linear filters are mathematically described in terms of their transfer function or, in layman's terms, frequency response. A transfer function can be decomposed as a combination of *first order* responses and *second order* responses (implemented as so-called biquad sections). These can be described according to their so-called pole and zero frequencies, which are complex numbers in the case of second-order responses.

First Order Filters

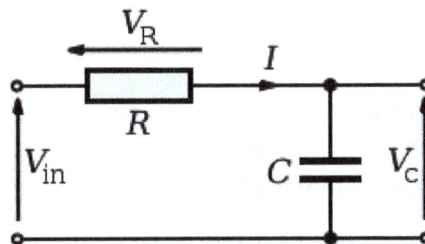

A first order low-pass (high-cut) filter implemented using only a resistor and capacitor.

A first order filter can alter the response of frequencies above and below a point. In the transition region the filter response will have a slope of up to 6 dB per octave. The bass and treble controls in a hi-fi system are each a first order filter in which the balance of frequencies above and below a point are varied using a single knob. A special case of first order filters is a first order high-pass or low-pass filter in which the 6 dB per octave cut of low or high frequencies extends indefinitely. These are the simplest of all filters to implement individually, requiring only a capacitor and resistor.

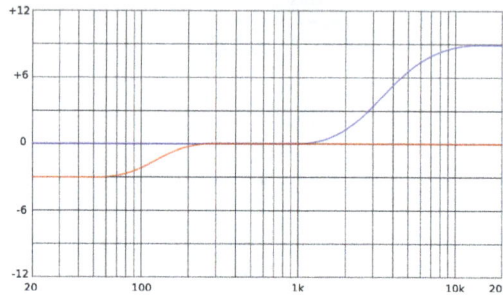

Two first-order shelving filters: a -3dB bass cut (red), and a +9dB treble boost (blue)

Second Order Filters

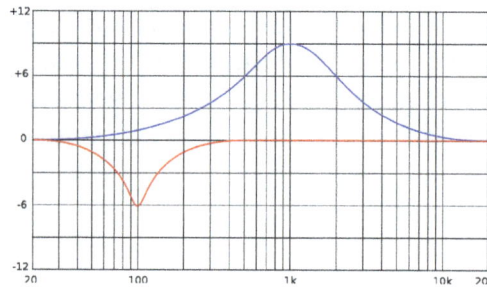

Second order filter responses

Second order filters are capable of resonance (or anti-resonance) around a particular frequency. The response of a second order filter is specified not only by its frequency but its Q; a higher Q corresponds to a sharper response (smaller bandwidth) around a particular center frequency. For instance, the red response in the accompanying image cuts frequencies around 100 Hz with a higher Q than the blue response which boosts frequencies around 1000 Hz. Higher Q's correspond to resonant behaviour in which the half-power or -3 dB bandwidth, BW, is given by:

$$BW = F_0 / Q$$

where F_0 is the resonant frequency of the second order filter. BW is the bandwidth expressed in the same frequency unit that F_0 is. Low Q filter responses (where $Q < \frac{1}{2}$) are not said to be resonant and the above formula for bandwidth does not apply.

It is also possible to define the Q of a band-pass function as:

$$Q = \frac{\sqrt{2^N}}{2^N - 1} = \frac{1}{2\sinh\left(\frac{\ln(2)}{2} N\right)},$$

where N is the bandwidth in octaves. The reverse mapping is

$$N = 2\log_2\left(\frac{1}{2Q} + \sqrt{\frac{1}{4Q^2} + 1}\right) = \frac{2}{\ln(2)}\operatorname{arcsinh}\left(\frac{1}{2Q}\right).$$

It should be noted that a second-order filter response with Q of less than 1/2 can be decomposed into two first-order filter functions, a low-cut and a high-cut (or boost). Of more interest are resonant filter functions which can boost (or cut) a narrow range of frequencies. In addition to specifying the center frequency F_0 and the Q, the specification of the filter's zeros determines how much that frequency band will be boosted (or cut). Thus a parametric equalizer section will have three controls for its center frequency F_0, bandwidth or Q, and the amount of boost or cut usually expressed in dB.

The range of second-order filter functions is important because any analog filter function can be decomposed into a (usually small) number of these (plus, perhaps, simpler first order responses). These are implemented directly by each section of a parametric equalizer where they are explicitly adjusted. And each element of a graphic equalizer based on a filter bank includes one such element whose Q is not adjustable by the user.

Uses

In sound recording, equalization is used to improve an instrument's sound or make certain instruments and sounds more prominent. For example, a recording engineer may use an equalizer to make some high-pitches in a vocal part louder while making low-pitches in a drum part quieter.

Equalization is commonly used to increase the 'depth' of a mix, creating the impression that some sounds in a mono or stereo mix are farther or closer than others, relatively. Equalization is also commonly used to give tracks with similar frequency components complementary spectral contours, known as mirrored equalization. Select components of parts which would otherwise compete, such as bass guitar and kick drum, are boosted in one part and cut in the other, and vice versa, so that they both stand out.

Equalizers can correct problems posed by a room's acoustics, as an auditorium will generally have an uneven frequency response especially due to standing waves and acoustic dampening. The frequency response of a room may be analyzed using a spectrum analyzer and a pink noise generator for instance. Then a graphic equalizer can be easily adjusted to compensate for the room's acoustics. Such compensation can also be applied to tweak the sound quality of a recording studio in addition to its use in live sound reinforcement systems and even home hi-fi systems.

During live events where signals from microphones are amplified and sent to speaker systems, equalization is not only used to "flatten" the frequency response but may also be useful in eliminating feedback. When the sound produced by the speakers is picked up by a microphone, it is further reamplified; this recirculation of sound can lead to "howling" requiring the sound technician to reduce the gain for that microphone, perhaps sacrificing the contribution of a singer's voice for instance. Even at a slightly reduced gain, the feedback will still cause an unpleasant resonant sound around the frequency at which it would howl. But because the feedback is troublesome at a particular frequency, it is possible to cut the gain only around that frequency while preserving the gain at most other frequencies. This can best be done using a parametric equalizer tuned to that very frequency with its amplitude control sharply reduced. By adjusting the equalizer for a narrow bandwidth (high Q), most other frequency components will not be affected. The extreme case when the signal at the filter's center frequency is completely eliminated is known as a notch filter.

An equalizer can be used to correct or modify the frequency response of a speaker system rather than designing the speaker itself to have the desired response. For instance, the Bose 901 speaker system doesn't use separate larger and smaller drivers to cover the bass and treble frequencies. Instead it uses nine drivers all of the same four-inch diameter, more akin to what one would find in a table radio. However this speaker system is sold with an active equalizer. That equalizer must be inserted into the amplifier system so that the amplified signal that is finally sent to the speakers has its response increased at the frequencies where the response of these drivers falls off, and vice versa, producing the response intended by the manufacturer.

Tone controls (usually designated "bass" and "treble") are simple shelving filters included in most hi-fi equipment for gross adjustment of the frequency balance. The bass control may be used, for instance, to increase the drum and bass parts at a dance party, or to reduce annoying bass sounds when listening to a person speaking. The treble control might be used to give the percussion a sharper or more "brilliant" sound, or can be used to cut such high frequencies when they have been overemphasized in the program material or simply to accommodate a listener's preference.

A "rumble filter" is a high pass (low cut) filter with a cutoff typically in the 20 to 40 Hz range; this is the low frequency end of human hearing. "Rumble" is a type of low frequency noise produced in record players and turntables, particularly older or low quality models. The rumble filter prevents this noise from being amplified and sent to the loudspeakers. Some cassette decks have a switchable "subsonic Filter" feature that does the same thing for recordings.

A crossover network is a system of filters designed to direct electrical energy separately to the woofer and tweeter of a 2-way speaker system (and also to the mid-range speaker of a 3-way system). This is most often built into the speaker enclosure and hidden from the user. However, in bi-amplification, these filters operate on the low level audio signals, sending the low and high frequency components to separate amplifiers which connect to the woofers and tweeters respectively.

Equalization is used in a reciprocal manner in certain communication channels and recording technologies. The original music is passed through a particular filter to alter its frequency balance, followed by the channel or recording process. At the end of the channel or when the recording is played, a complementary filter is inserted which precisely compensates for the original filter and recovers the original waveform. For instance, FM broadcast uses a pre-emphasis filter to boost the high frequencies before transmission, and every receiver includes a matching de-emphasis filter to restore it. The white noise that is introduced by the radio is then also de-emphasized at the higher frequencies (where it is most noticeable) along with the pre-emphasized program, making the noise less audible. Tape recorders used the same trick to reduce "tape hiss" while maintaining fidelity. On the other hand, in the production of vinyl records, a filter is used to reduce the amplitude of low frequencies which otherwise produce large amplitudes on the tracks of a record. Then the groove can take up less physical space, fitting more music on the record. The preamp attached to the phono cartridge has a complementary filter boosting those low frequencies following the standard RIAA equalization curve.

Alignment Level

The Alignment level in an audio signal chain or on an audio recording is an anchor point that represents a reasonable or typical level. It does not represent a particular sound level or signal level or digital representation, but it can be defined as corresponding to particular levels in each of these domains.

For example, alignment level is commonly 0dBu in broadcast chains in places where the signal exists as analogue voltage. It most commonly is at -18dB FS (18dB below full scale digital) on digital recordings for programme exchange, in accordance with European Broadcasting Union (EBU) recommendations. 24-bit original or master recordings commonly have alignment level at -24dB FS in order to allow extra headroom, which can then be reduced to match the available headroom of the final medium by Audio level compression. FM broadcasts usually have only 9dB of headroom as recommended by the EBU.

Using alignment level rather than maximum permitted level as the reference point allows more sensible headroom management throughout the audio chain, so that quality is only sacrificed through level compression as late as possible.

The Reason for Alignment Level

Using alignment level rather than maximum permitted level as the reference point allows more sensible headroom management throughout the audio chain, so that quality is only sacrificed through compression as late as possible.

Loudness wars have caused a general fall in audio quality, initially on radio stations and more recently on CDs. As radio stations competed for attention and to raise the listener scores their ad revenue is based on, they used audio compression to give their sound more impact. They used level compressors, and in particular multi-band compressors that compress different frequencies independently. Such compressors usually incorporate fast acting limiters to eliminate brief peaks, since brief peaks, though they may not contribute much to perceived loudness, limit the modulation level that can be applied to FM transmissions in particular, if serious clipping and distortion are to be avoided. Digital broadcasting has changed all this: stations are no longer found by tuning across the band, so the loudest stations no longer stand out. Low noise level is also guaranteed regardless of signal level, so that it is no longer necessary to fully modulate to ensure acceptable clarity in poor reception areas. Many professionals feel that the more widespread adoption and understanding of alignment level throughout the audio industry could help bring modulation levels down, leaving headroom to cope with brief peaks, and using a different form of level compression that reduces dynamic range on programmes where this is considered desirable, but does not remove the brief peaks which add 'sparkle' and contribute to clearer sound. CDs in particular have suffered a loss of quality since they were introduced through the widespread use of fast limiting, which, given their very low noise level is quite unnecessary.

Digital audio players such as the iPod, demonstrate the need for a common alignment level. While tracks taken from recent CDs sound loud enough, many older recordings (such as Pink Floyd albums which notably allowed lots of headroom for stunning dynamic range and rarely reach peak digital level) are far too quiet, even at full volume setting. Older audio systems typically incorporated 12dB of 'overvolume', meaning that it was possible to turn up the loudness on a quiet recording to make maximum use of amplifier output even if peak level was never reached on the recording. Modern devices, however, tend to produce maximum output at full volume only on recordings that reach full-scale digital level. If extra gain is added, then playing a modern CD after listening to a well recorded older one is likely to deafen, requiring the volume control to be turned down by a huge amount. Again, the adoption of a common alignment level (early CDs allowed around 18dB of headroom by common consent) would make sense, improving quality and usability and ending the loudness war.

Making Compression a Listening Option

The incorporation of (switchable) level compression in domestic music systems and car in-car systems would allow higher quality on systems capable of wide dynamic range and in situations that allowed realistic reproduction. Such compression systems have been suggested and tried from time to time, but are not in widespread use — a 'chicken and egg' problem since producers feel they must make programmes and recordings that sound good in car with high ambient noise or on cheap low-power music systems.

In the UK, some DAB receivers do incorporate a menu setting for automatic loudness compensation which adds extra gain on BBC Radio 3 and BBC Radio 4, to allow for the fact that these programmes adopt lower levels than, for example, the pop station Radio 1. Some television receivers also have a menu setting for loudness normalisation, aimed at helping to reduce excessive loudness on advertisements. However, there is no common agreement to reduce compression and limiting and leave these tasks to the receiver.

Audio Levels

Analogue audio signal levels have historically been categorised for professional usage in two groups – microphone level (low) and line level (high). Domestic equipment generally falls into a third category, just below professional line level. Signal levels are measured using units based on the Decibel. There are many different types of Level Meter used for signal measurement and monitoring.

Absolute Levels

Decibels specify ratios, not absolute levels. To specify an absolute level, we state the ratio of the quantity to be specified referred to a standard reference value. For example, signal power in electronic circuits is usually referred to one milliwatt. The level would then be given as so many decibels re: 1 mW.

In all formal writing, and wherever there could possibly be any doubt, it is essential to give the reference quantity. This may be done by putting the reference quantity in brackets. For example, where the reference is 1 mW, the level would be given as so many dB(mW).

The abbreviation dB is commonly modified by adding an extra letter to indicate the reference value: for example dBm indicates that the reference level is 1 mW. Such abbreviations are not recommended by the AES and IEC but are likely to remain in use for some time.

The abbreviations dBA and dB(A) do not indicate a reference value but a frequency weighting.

dBm

In the early days of broadcast audio, level standards were derived from the telephone industry. At that time, the standard signal level was referred to as 0 dBm, and this was the amount of signal required to dissipate 1 milliwatt of power in a 600 ohm termination (600 ohms send and receive impedances in equipment was also a standard of the time).

The 'm' in 0 dBm indicates that the measurement was made to this standard (ie. 600

ohm terminations in place). If you do the sums, you will find that the voltage needed across a 600 ohm load to produce the 1 mW of power required is 0.775 of a volt.

dBu

Today, we generally don't use 600 ohm impedances any more in professional audio since we are not interested in the efficient transmission of power. Instead, we use low output impedances and high input impedances (typically 100 ohms or less, and 10,000 ohms or more, respectively). However we still use the same signal voltage as the original dBm reference level – ie. 0.775 V.

To differentiate between this new standard and the old (matched 600 ohm impedance) one, we now use the suffix 'u' – hence the dBu. In essence, then, the dBu means that our reference signal is 0.775 V irrespective of impedance.

The standard Alignment Level employed by UK broadcasters is 0 dBu, shown on peak programme meters as PPM4, and on standard VUs as 0VU (or -4VU in some establishments). However +4 dBu is a common operating level in many music studios and professional equipment is often aligned to this standard.

dBV

Domestic/consumer equipment is usually built with a much lower reference signal level which is normally defined as -10 dBV. In this case the V indicates that the reference signal is 1 Volt. Again, if you do the sums you discover that -10 dBV implies a signal of 0.316 Volts. If you convert this into (professional) dBu terms, it equates to -7.78 dBu.

In other words, the output of a consumer device working to the -10 dBV standard is going to be roughly 8 dB or 12 dB below the nominal level expected of a professional systems.

dBFS

In digital equipment, the only defining point is the level at which you run out of quantising levels – the digital overload point. All digital equipment defines this as 0 dBFS, where FS stands for Full Scale.

For practical usage, we need to define a working headroom to allow for normal and unexpected peak levels (bearing in mind that PPMs deliberately don't show fast transients), and so various standards bodies have specified alternative Alignment Levels between -20 and -12 dBFS.

In Europe, the EBU recommend that -18 dBFS equates to the Alignment Level, and for UK broadcasters, Alignment Level is taken as 0 dBu (PPM4 or -4VU). This works well with original recordings and allows a reasonable safety margin against overloads. In a post-production environment, where levels are likely to be much better controlled (through manual or automatic means), many organisations use a higher reference level. One common alternative is to use -12 dBFS which takes advantage of the fact that less 'contingency' headroom is required, and trades that for an improvement in the noise performance of the medium.

Another approach, often taken with systems offering 24 bit wordlengths, is to provide an increased headroom margin while maintaining the same (or better) noise performance of a 16 bit system. Typically, people are using -20 or -22 dBFS in this situation. The American SMPTE standard defines -20 dBFS as the Alignment Level for all systems anyway, regardless of digital wordlength.

LUFS

In August 2010 the EBU made Recommendation 128 in respect of "Loudness normalisation and permitted maximum level of audio signals", introducing the term LUFS (Loudness Unit, referenced to Full Scale).

This states that "In addition to the average loudness of a programme ('Programme Loudness') the EBU recommends that the descriptors 'Loudness Range' and 'Maximum True Peak Level' be used for the normalisation of audio signals, and to comply with the technical limits of the complete signal chain as well as the aesthetic needs of each programme/station depending on the genre(s) and the target audience."

References

- Peek, Hans; Bergmans, Jan; Van Haaren, Jos; Toolenaar, Frank; Stan, Sorin (2009). Origins and Successors of the Compact Disc (Philips Research Book Series, Volume 11). Springer Science+Business Media B.V. p. 10. ISBN 978-1-4020-9552-8

- Janssens, Jelle; Stijn Vandaele; Tom Vander Beken (2009). "The Music Industry on (the) Line? Surviving Music Piracy in a Digital Era". European Journal of Crime. 77 (96). doi:10.1163/157181709X429105

- "Extraordinary EQ from Extraordinary Engineers". Mackie. Archived from the original on December 2, 2013. Retrieved 2013-11-25

- Peterson, George; Robair, Gino [ed.] (1999). Alesis ADAT: The Evolution of a Revolution. Mixbooks. p. 2. ISBN 0-87288-686-7

- Fine, Thomas (2008). Barry R. Ashpole, ed. "The Dawn of Commercial Digital Recording" (PDF). ARSC Journal. Ted P. Sheldon. Retrieved 2010-05-02

2

Understanding Audio Processing

Audio signal processing is the alteration of audio signals by using an effects unit or an audio effect. This is an introductory chapter, which will introduce briefly all the significant aspects of audio processing. It includes a number of significant topics such as dynamic range compression, audio mastering, audio mixing, audio post production, sound effect, etc.

Audio Processing

The two principal human senses are vision and hearing. Correspondingly, much of DSP (Digital Signal Processing) is related to image and audio processing. People listen to both music and speech. DSP has made revolutionary changes in both these areas.

The path leading from the musician's microphone to the audiophile's speaker is remarkably long. Digital data representation is important to prevent the degradation commonly associated with analog storage and manipulation. This is very familiar to anyone who has compared the musical quality of cassette tapes with compact disks. In a typical scenario, a musical piece is recorded in a sound studio on multiple channels or tracks. In some cases, this even involves recording individual instruments and singers separately. This is done to give the sound engineer greater flexibility in creating the final product. The complex process of combining the individual tracks into a final product is called mix down. DSP can provide several important functions during mix down, including: filtering, signal addition and subtraction, signal editing, etc.

Audio Processing means changing the characteristics of an audio signal in some way. Processing can be used to enhance audio, fix problems, separate sources, create new sounds, as well as to compress, store and transmit data.

Audio signal processing is at the heart of recording, enhancing, storing and transmitting audio content. Audio signal processing is used to convert between analog and digital formats, to cut or boost selected frequency ranges, to remove unwanted noise, to add effects and to obtain many other desired results. Today, this process can be done on an ordinary PC or laptop, as well as specialized recording equipment.

Analog Signals

Analog indicates something that is mathematically represented by a continuous function. Thus, an analog signal is one represented by a continuous stream of data, in this case along an electrical circuit in the form of voltage or current. Analog signal processing then involves physically altering the continuous signal by changing the voltage or current or charge via various electrical means.

Historically, before the advent of widespread digital technology, analog was the only method by which to manipulate a signal. Since that time, as computers and software have become more capable and affordable and digital signal processing has become the method of choice.

Digital Signals

A digital representation expresses the audio waveform as a sequence of symbols, usually binary numbers. This permits signal processing using digital circuits such as digital signal processors, microprocessors and general-purpose computers. Most modern audio systems use a digital approach as the techniques of digital signal processing are much more powerful and efficient than analog domain signal processing.

Application Areas

Processing methods and application areas include storage, level compression, data compression, transmission, enhancement (e.g., equalization, filtering, noise cancellation, echo or reverb removal or addition, etc.).

Audio Broadcasting

Arguably the most important audio processing in audio broadcasting takes place just before the transmitter. The audio processor here must

- prevent or minimize overmodulation,

- compensate for non-linear transmitters (a potential issue with medium wave and shortwave broadcasting) and

- adjust overall loudness to desired level.

Echo

Sound is a mechanical wave which travels through a medium from one location to another. This motion through a medium occurs as one particle of the medium interacts with its neighboring particle, transmitting the mechanical motion and corresponding energy to it. This transport of mechanical energy through a medium by particle interaction is what makes a sound wave a mechanical wave.

An echo is a sound that is repeated because the sound waves are reflected back. Sound waves can bounce off smooth, hard objects in the same way as a rubber ball bounces off the ground. Although the direction of the sound changes, the echo sounds the same as the original sound. Echoes can be heard in small spaces with hard walls, like wells, or where there are lots of hard surfaces all around. That is why echoes can be heard in a canyon, cave, or mountain range. But sounds are not always reflected. If they meet a soft surface, such as a cushion, they will be absorbed and will not bounce back.

Conditions Necessary for Hearing the Echo

The distance between the sound source and the reflecting surface must not be less than 17 metres where the time period between hearing the original sound and its echo should not be less than 0.1 of a second.

The human ear can not distinguish between two successive sounds if the period between them is less than 0.1 second, Wide and big reflecting surface must be presented to hear the echo such as the walls, the mountains or the water bodies.

The velocity of sound through the air is 340 m / sec, so the distance travelled by the sound and its echo in 0.1 sec = 34 metres, The distance travelled by the sound is 17 metres from the sound source to the reflecting surface and 17 metres from the reflecting surface to the ear (echo).

The echo can not be heared if the distance between the sound source and the reflecting surface is less than 17 metres because the time between hearing the main sound and its echo will be less than 0.1 of a second, and the human ear can not distinguish between the two successive sounds.

When the distance between hearing the main sound and the reflecting surface is multiplies of 17 metres (twice or three times), the echo is heard in the form of the last two or three phrases of the whole produced sound.

Acoustic Phenomenon

Acoustic waves are reflected by walls or other hard surfaces, such as mountains and privacy fences. The reason of reflection may be explained as a discontinuity in the propagation medium. This can be heard when the reflection returns with sufficient magnitude and delay to be perceived distinctly. When sound, or the echo itself, is reflected multiple times from multiple surfaces, the echo is characterized as a reverberation.

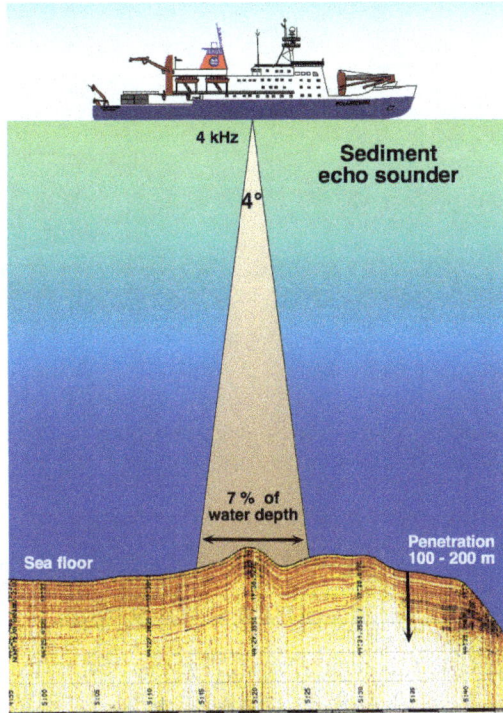

This illustration depicts the principle of sediment echo sounding, which uses
a narrow beam of high energy and low frequency

The human ear cannot distinguish echo from the original direct sound if the delay is
less than 1/10 of a second. The velocity of sound in dry air is approximately 343 m/s at
a temperature of 25 °C. Therefore, the reflecting object must be more than 17.2m from
the sound source for echo to be perceived by a person located at the source. When a
sound produces an echo in two seconds, the reflecting object is 343m away. In nature,
canyon walls or rock clliffs facing water are the most common natural settings for hearing echoes. The strength of echo is frequently measured in dB sound pressure level
(SPL) relative to the directly transmitted wave. Echoes may be desirable (as in sonar)
or undesirable (as in telephone systems).

In Music

In music performance and recording, electric echo effects have been used since the
1950s. The Echoplex is a tape delay effect, first made in 1959 that recreates the sound
of an acoustic echo. Designed by Mike Battle, the Echoplex set a standard for the effect
in the 1960s and was used by most of the notable guitar players of the era; original
Echoplexes are highly sought after. While Echoplexes were used heavily by guitar players (and the occasional bass player, such as Chuck Rainey, or trumpeter, such as Don
Ellis), many recording studios also used the Echoplex. Beginning in the 1970s, Market
built the solid-state Echoplex for Maestro. In the 2000s, most echo effects units use
electronic or digital circuitry to recreate the echo effect.

Famous Echoes

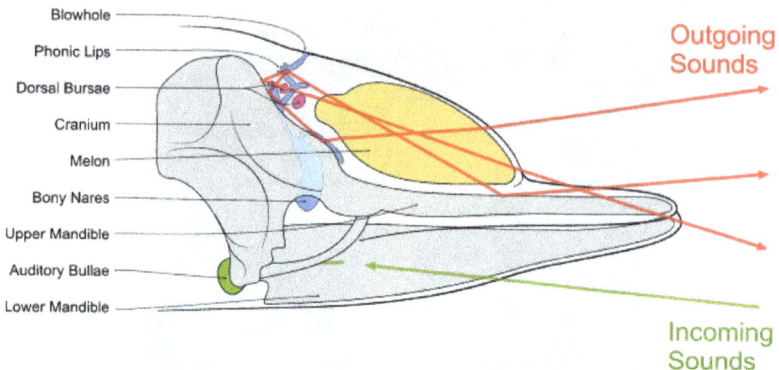

Whales echolocation organs, which produce echoes and receive sounds.
Arrows illustrate the outgoing and incoming path of sound.

- Inchindown oil tanks, current record holder for longest echo.

- Hamilton Mausoleum, Hamilton, South Lanarkshire, Scotland: Its high stone means it takes 15 seconds for the sound of a slammed door to delay.

- Gol Gumbaz of Bijapur, India: Any whisper, clap or sound gets echoed repeatedly.

- The Golkonda Fort of Hyderabad, India.

- The Echo Wall at the Temple of Heaven, Beijing, China.

- The Whispering Gallery of St Paul's Cathedral, London, England, UK.

- Echo Point, the Three Sisters, Katoomba, Australia.

- The Temple of Kukulcan *("El Castillo")*, Chichen Itza, Mexico.

- The Baptistry of Pisa, Pisa, Italy.

- The echo near Milan visited by Mark Twain in *The Innocents Abroad*.

- The echo in Chinon, France which is used in a traditional local rhyme.

- The gazebo of Napier Museum in Trivandrum, Kerala, India.

Dynamic Range Compression

Dynamic range compression, despite being one of the most widely used audio effects, is still poorly understood, and there is little formal knowledge and analysis of compressor design techniques.

Dynamic Range Compression (DRC) is the process of mapping the dynamic range of an audio signal to a smaller range, i.e., reducing the signal level of the higher peaks while leaving the quieter parts untreated. DRC is used extensively in audio recording, production work, noise reduction, broadcasting, and live performance applications.

Considering the classic audio effects (equalization, delay, panning, etc.), the dynamic range compressor is perhaps the most complex one. Design choices involve the compressor topology, the static compression characteristic, placement, and type of smoothing filters, sidechain filtering, etc. This explains why every compressor in common usage behaves and sounds slightly different and why certain compressor models have become audio engineers' favorites for certain types of signal. The analysis of compressor designs is difficult because they represent nonlinear time-dependent systems with memory. The gain reduction is applied smoothly and not instantaneously as would be the case with a simple static nonlinearity. Furthermore the large number of design choices makes it nearly impossible to draw a generic compressor block diagram that would be valid for the majority of real world compressors. "No two compressors sound alike, each one is inaccurate in its own unique way." Some differ in topology, others introduce additional stages, and some simply differ from the precise digital design since these deviations add character to the compressor. However we can describe the main parameters of a compressor unit and specify a set of standard stages and building blocks that are present in almost any compressor design.

A dedicated electronic hardware unit or audio software that applies compression is called a compressor. In the 2000s, compressors became available as software plugins that run in digital audio workstation software. In recorded and live music, compression parameters may be adjusted to change the way they affect sounds. Compression and limiting are identical in process but different in degree and perceived effect. A limiter is a compressor with a high ratio and, generally, a fast attack time.

Types

Two Methods of Dynamic Range Compression

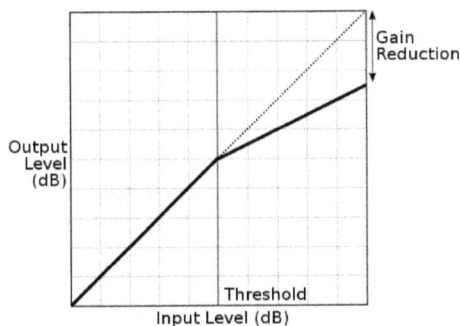

Downward compression   Upward compression

Downward compression reduces loud sounds over a certain threshold while quiet sounds remain unaffected. A limiter is an extreme type of downward compression.

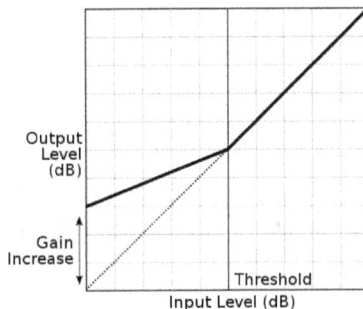

Upward compression increases the loudness of sounds below a certain threshold while leaving louder sounds unaffected. Both downward and upward compression reduce the dynamic range of an audio signal.

An expander increases the dynamic range of the audio signal. Expanders are generally used to make quiet sounds even quieter by reducing the level of an audio signal that falls below a set threshold level. A noise gate is a type of expander.

Design

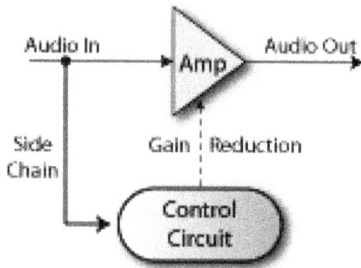

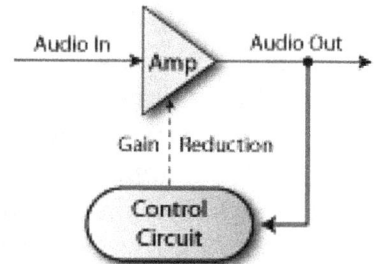

A feed-forward compressor design (left) and feedback design (right)

The signal entering a compressor is split; one copy is sent to a variable-gain amplifier and the other to a *side-chain* where the signal level is measured and a circuit controlled by the measured signal level applies the required gain to the amplifier. This design, known as a *feed-forward* type, is used today in most compressors. Earlier designs were based on a *feedback* layout where the signal level was measured after the amplifier.

There are a number of technologies used for variable-gain amplification, each having different advantages and disadvantages. Vacuum tubes are used in a configuration called *variable-mu* where the grid-to-cathode voltage changes to alter the gain. Optical compressors use a photoresistor and a small lamp (Incandescent, LED or electroluminescent panel) to create changes in signal gain. Other technologies used include field effect transistors and a diode bridge.

When working with digital audio, digital signal processing techniques are commonly used to implement compression as audio plug-ins, in mixing consoles, and in digital audio workstations. Often the algorithms used emulate the above analog technologies.

Controls and Features

A number of user-adjustable control parameters and features are used to adjust dynamic range compression signal processing algorithms and components.

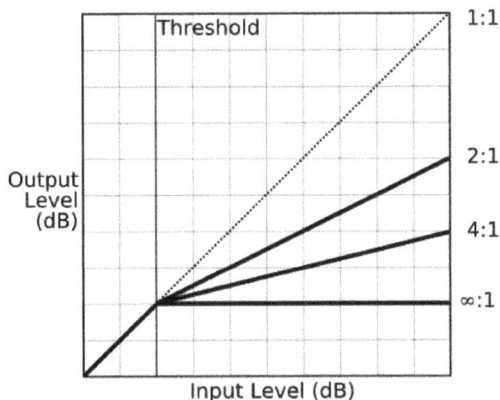

Different compression ratios for a signal level above the theshold

Threshold

A compressor reduces the level of an audio signal if its amplitude exceeds a certain *threshold*. Threshold is commonly set in decibels dB, where a lower threshold (e.g. -60 dB) means a larger portion of the signal is treated. When the signal level is below the threshold, no processing is performed and the input signal is passed, unmodified, to the output. Thus a higher threshold of, e.g., −5 dB, results in less processing, less compression.

Threshold timing behavior is subject to attack and release settings. When the signal level goes above threshold, compressor operation is delayed by the *attack* setting. For an amount of time determined by the *release* after the input signal has fallen below the threshold, the compressor continues to apply dynamic range compression.

Ratio

The amount of gain reduction is determined by ratio: a ratio of 4:1 means that if input level is 4 dB over the threshold, the output signal level is reduced to 1 dB over the threshold. The gain and output level has been reduced by 3 dB.

The highest ratio of ∞:1 is often known as *limiting*. It is commonly achieved using a ratio of 60:1, and effectively denotes that any signal above the threshold is brought down to the threshold level once the *attack* time has expired.

Attack and Release

A compressor may provide a degree of control over how quickly it acts. The *attack* is the period when the compressor is decreasing gain in response to increased level at the input to reach the gain determined by the ratio. The *release* is the period when the compressor is increasing gain in response to reduced level at the input to reach the output gain determined by the ratio, or, to unity, once the input level has fallen below the threshold.

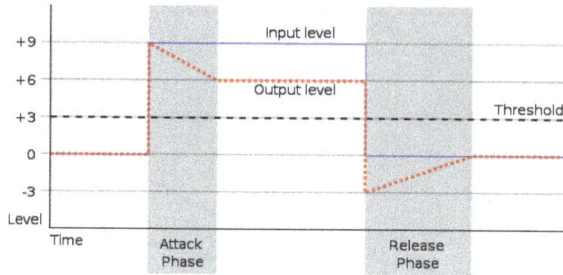

The attack and release phases in a compressor

The length of each period is determined by the rate of change and the required change in gain. For more intuitive operation, a compressor's attack and release controls are labeled as a unit of time (often milliseconds). This is the amount of time it takes for the gain to change a set amount of dB, decided by the manufacturer—often 10 dB. For example, if the compressor's time constants are referenced to 10 dB, and the attack time is set to 1 ms, it takes 1 ms for the gain to decrease by 10 dB, and 2 ms to decrease by 20 dB.

In many compressors the attack and release times are adjustable by the user. Some compressors, however, have the attack and release times determined by the circuit design and these cannot be adjusted by the user. Sometimes the attack and release times are 'automatic' or 'program dependent', meaning that the times change depending on the input signal. Because the loudness pattern of the source material is modified by the compressor it may change the character of the signal in subtle to quite noticeable ways depending on the settings used.

Soft and Hard Knees

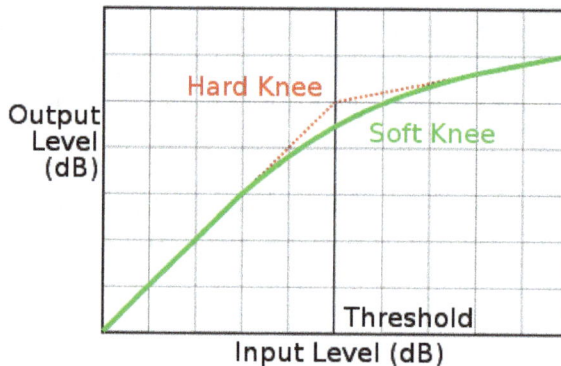

Hard Knee and Soft Knee compression

Another control a compressor might offer is hard/soft knee. This controls whether the bend in the response curve between below threshold and above threshold is abrupt (hard) or gradual (soft). A soft knee slowly increases the compression ratio as the level increases and eventually reaches the compression ratio set by the user. A soft knee reduces the audible change from uncompressed to compressed, especially for higher ratios where a hard knee changeover would be more noticeable.

Peak vs RMS Sensing

A peak-sensing compressor responds to the instantaneous level of the input signal. While providing tighter peak control, peak sensing might yield very quick changes in gain reduction, more evident compression or sometimes even distortion. Some compressors apply an averaging function (commonly root mean square or RMS) on the input signal before comparing its level to the threshold. This produces a more relaxed compression that more closely relates to human perception of loudness.

Stereo Linking

A compressor in stereo linking mode applies the same amount of gain reduction to both the left and right channels. This is done to prevent image shifting that can occur if each channel is compressed individually. It becomes noticeable when a loud element that is panned to either edge of the stereo field raises the level of the program to the compressor's threshold, causing its image to shift toward the center of the stereo field.

Stereo linking can be achieved in two ways: Either the compressor sums to mono the left and right channel at the input, then only the left channel controls are functional; or, the compressor still calculates the required amount of gain reduction independently for each channel and then applies the highest amount of gain reduction to both (in such case it could still make sense to dial different settings on left and right channels as one might wish to have less compression for left-side events).

Makeup Gain

Because the compressor is reducing the gain (or level) of the signal, the ability to add a fixed amount of *make-up gain* at the output is usually provided so that an optimum level can be used.

Look-ahead

The look-ahead function is designed to overcome the problem of being forced to compromise between slow attack rates that produce smooth-sounding gain changes, and fast attack rates capable of catching transients. Look-ahead is a misnomer in that the future is not actually observed. Instead, the input signal is split, and one side is delayed. The non-delayed signal is used to drive the compression of the delayed signal, which then appears at the output. This way a smooth-sounding slower attack rate can be used to catch transients. The cost of this solution is that the signal is delayed.

Uses

Public Spaces

Compression is often used to make music sound louder without increasing its peak amplitude. By compressing the peak (or loudest) signals, it becomes possible to increase the overall gain (or volume) of a signal without exceeding the dynamic limits of a reproduction device or medium. The net effect, when compression is applied along with a gain boost, is that relatively quiet sounds become louder, while louder sounds remain unchanged.

Compression is often applied in this manner in audio systems for restaurants, retail, and similar public environments that play background music at a relatively low volume and need it compressed, not just to keep the volume fairly constant, but also to make quiet parts of the music audible over ambient noise.

Compression can increase average output gain of a power amplifier by 50 to 100% with a reduced dynamic range. For paging and evacuation systems, this adds clarity under noisy circumstances and saves on the number of amplifiers required.

Music Production

Compression is often used in music production to make performances more consistent in dynamic range so that they "sit" in the mix of other instruments better and maintain consistent attention from the listener. Vocal performances in rock music or pop music are usually compressed to make them stand out from the surrounding instruments and add clarity.

Compression can also be used on instrument sounds to create effects not primarily focused on boosting loudness. For instance, drum and cymbal sounds tend to decay quickly, but a compressor can make the sound appear to have a more sustained tail. Guitar sounds are often compressed to produce a fuller, more sustained sound.

Most devices capable of compressing audio dynamics can also be used to reduce the volume of one audio source when another audio source reaches a certain level; this is called side-chaining.

In electronic dance music, side-chaining is often used on basslines, controlled by the kick drum or a similar percussive trigger, to prevent the two conflicting, and provide a pulsating, rhythmic dynamic to the sound.

Voice

A compressor can be used to reduce sibilance ('ess' sounds) in vocals by feeding the compressor with an EQ set to the relevant frequencies, so that only those frequencies activate the compressor. If unchecked, sibilance could cause distortion even if sound levels are not very high. This usage is called de-essing.

Compression is used in voice communications in amateur radio that employ SSB modulation. Often it is used to make a particular station's signal more readable to a distant station, or to make one's station's transmitted signal stand out against others. This occurs especially in pileups where amateur radio stations are competing for the opportunity to talk to a DX station. Since an SSB signal's amplitude depends on the level of modulation, the net result is that the average amplitude of the signal, and hence average transmitted power is stronger than it would be without compression. Most modern amateur radio SSB transceivers have speech compressors built in.

Compression is also used in land mobile radio, especially in transmit audio of professional walkie-talkies and in remote control dispatch consoles.

Broadcasting

Compression is used extensively in broadcasting to boost the perceived volume of sound while reducing the dynamic range of source audio (typically CDs) to a range that can be accommodated by the narrower-range broadcast signal. Broadcasters in most countries have legal limits on instantaneous peak volume they may broadcast. Normally these limits are met by permanently inserted hardware in the on-air chain.

As was alluded to above, the use of compressors to boost perceived volume is a favorite trick of broadcasters who want their station to sound "louder" at the same volume than comparable stations on the dial. The effect is to make the more heavily compressed station "jump out" at the listener at a given volume setting.

But loudness jumps are not limited to inter-channel differences; they also exist between programme material within the same channel. Loudness differences are a frequent source of audience complaints, especially TV commercials and promos that seem too loud. Complicating this is that many broadcasters use (quasi-)peak meters and peak-levelling. Unfortunately the peak level reading does not correlate well with the perceived loudness. It basically should only be used to prevent overmodulation.

The European Broadcasting Union has been addressing this issue in the EBU PLOUD Group, which consist of over 240 audio professionals, many from broadcasters and equipment manufacturers. In 2010, the EBU published EBU Recommendation R 128, which introduces a new way of metering and normalising audio. The Recommendation is based on ITU-R BS.1770. Several European TV stations have already announced to support the new norm and over 20 manufacturers have announced products supporting the new 'EBU Mode' Loudness meters.

To help audio engineers understand what Loudness Range their material consists of (e.g. to check if some compression may be needed to fit it into the channel of a specific delivery platform), the EBU also introduced the Loudness Range Descriptor.

Marketing

Record companies, mixing engineers and mastering engineers have been gradually increasing the overall volume of commercial albums. The greater loudness is achieved by using higher degrees of compression and limiting during mixing and mastering; compression algorithms have been engineered specifically to accomplish the task of maximizing audio level in the digital stream. Hard limiting or clipping can result, affecting the tone and timbre of the music. The effort to increase loudness has been referred to as the loudness war.

Most television commercials are heavily compressed (typically to a dynamic range of no more than 3 dB) to achieve near-maximum perceived loudness while staying within permissible limits. This causes a problem that TV viewers often notice: when a station switches from minimally compressed program material to a heavily compressed commercial, the volume sometimes seems to increase dramatically. Peak loudness might be the same—meeting the letter of the law—but high compression puts much more of the audio in the commercial at close to the maximum allowable, making the commercial seem much louder.

Other uses

Some applications use a compressor to reduce the dynamic range of a signal for transmission, expanding it afterward. This reduces the effects of a channel with limited dynamic range.

Bass amplifiers and keyboard amplifiers often include compression circuitry to prevent sudden high-wattage peaks that could damage the speakers. Electric bass players often use compression effects, either effects units available in pedal, rackmount units, or built-in devices in bass amps, to even out the sound levels of their basslines.

Gain pumping, where a regular amplitude peak (such as a kick drum) causes the rest of the mix to change in volume due to the compressor, is generally avoided in music production. However, many dance and hip-hop musicians purposefully use this phenomenon, causing the mix to alter in volume rhythmically in time with the beat.

Hearing aids use a compressor to bring the audio volume into the listener's hearing range. To help the patient perceive the direction sound comes from, some hearing aids use binaural compression.

Compressors are also used for hearing protection in some electronic "active sound protection" earmuffs and earplugs, to let sounds at ordinary volumes be heard normally while attenuating louder sounds, possibly also amplifying softer sounds. This allows, for example, shooters wearing hearing protection at a shooting range to converse normally, while sharply attenuating the much louder sounds of the gunshots, and similarly for musicians to hear quiet music but be protected from loud noises such as drums or cymbal crashes.

Limiting

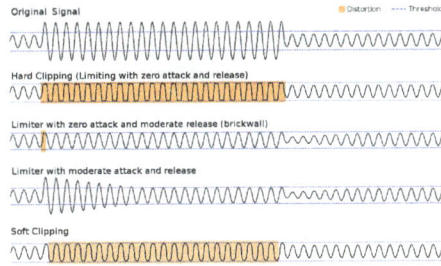

Original Signal

Hard Clipping (Limiting with zero attack and release)

Limiter with zero attack and moderate release (brickwall)

Limiter with moderate attack and release

Soft Clipping

Limiting and clipping compared. Note that clipping introduces a large amount of distortion whereas limiting only introduces a small amount while keeping the signal within the threshold.

Compression and limiting are identical in process but different in degree and perceived effect. A limiter is a compressor with a high ratio and, generally, a fast attack time. Compression with ratio of 10:1 or more is generally considered limiting.

Brick wall limiting has a very high ratio and a very fast attack time. Ideally, this ensures that an audio signal never exceeds the amplitude of the threshold. Ratios of 20:1 all the way up to ∞:1 are considered 'brick wall'. The sonic results of more than momentary and infrequent hard/brick-wall limiting are harsh and unpleasant, thus it is more common as a safety device in live sound and broadcast applications.

Some bass amps and PA system amplifiers include limiters to prevent sudden volume peaks from causing distortion and/or damaging the speakers. Some modern consumer electronics devices incorporate limiters. Sony uses the automatic volume limiter system (AVLS), on some audio products and the PlayStation Portable.

Side-chaining

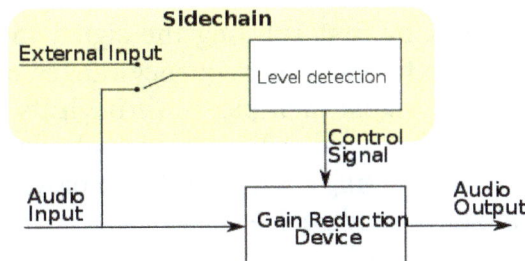

The sidechain of a feed-forward compressor

A compressor with a side-chain input controls gain from main input to output based on the level of the signal at the side-chain input. The compressor behaves in the conventional manner when both inputs are supplied with the same signal. The side-chain input is used by disc jockeys for ducking – lowering the music volume automatically when speaking. The DJ's microphone signal is routed to the side-chain input so that whenever the DJ speaks the compressor reduces the volume of the music. A sidechain with equalization controls can be used to reduce the volume of signals that have a strong spectral content within a certain frequency range: it can act as a de-esser, reducing the level of vocal sibilance in the range of 6–9 kHz. A de-esser helps reduce high frequencies that tend to overdrive preemphasized media (such as phonograph records and FM radio). Another use of the side-chain in music production serves to maintain a loud bass track without the bass drum causing undue peaks that result in loss of overall headroom.

Parallel Compression

One technique is to insert the compressor in a parallel signal path. This is known as parallel compression, a form of upward compression that facilitates dynamic control without significant audible side effects, if the ratio is relatively low and the compressor's sound is relatively neutral. On the other hand, a high compression ratio with significant audible artifacts can be chosen in one of the two parallel signal paths—this is used by some concert mixers and recording engineers as an artistic effect called *New York compression* or *Motown compression*. Combining a linear signal with a compressor and then reducing the output gain of the compression chain results in low-level detail enhancement without any peak reduction (since the compressor significantly adds to the combined gain at low levels only). This is often beneficial when compressing transient content, since it maintains high-level dynamic liveliness, despite reducing the overall dynamic range.

Multiband Compression

Multiband compressors can act differently on different frequency bands. The advantage of multiband compression over full-bandwidth compression is that unneeded audible gain changes or "pumping" in other frequency bands is not caused by changing signal levels in a single frequency band.

Multiband compressors work by first splitting the signal through some number of band-pass filters or crossover filters. The frequency ranges or crossover frequencies may be adjustable. Each split signal then passes through its own compressor and is independently adjustable for threshold, ratio, attack, and release. The signals are then recombined and an additional limiting circuit may be employed to ensure that the combined effects do not create unwanted peak levels.

Software plug-ins or DSP emulations of multiband compressors can be complex, with many bands, and require corresponding computing power.

Multiband compressors are primarily an audio mastering tool, but their inclusion in digital audio workstation plug-in sets is increasing their use among mix engineers. On-air signal chains of radio stations commonly use hardware multiband compressors to increase apparent loudness without fear of overmodulation. Having a louder sound is often considered an advantage in commercial competition. However, adjusting a radio station's multiband output compressor requires some artistic sense of style, plenty of time, and good ears. This is because the constantly changing spectral balance between audio bands may have an equalizing effect on the output, by dynamically modifying the on-air frequency response. A further development of this approach is programmable radio output processing, where the parameters of the multiband compressor automatically change between different settings according to the current programme block style or the time of day.

Serial Compression

Serial compression is a technique used in sound recording and mixing. Serial compression is achieved by using two fairly different compressors in a signal chain. One compressor generally stabilizes the dynamic range while the other aggressively compresses stronger peaks. This is the normal internal signal routing in common combination devices marketed as *compressor-limiters*, where an RMS compressor (for general gain control) is followed by a fast peak sensing limiter (for overload protection). Done properly, even heavy serial compression can sound natural in a way not possible with a single compressor. It is most often used to even out erratic vocals and guitars.

Software Audio Players

Some software audio players support plugins that implement compression. These can increase perceived volume of audio tracks, or even out the volume of highly-variable music (such as classical music, or a playlist that spans multiple music types). This improves listenability of audio played through poor-quality speakers, or when played in noisy environments (such as in a car or during a party). Such software may also be used in micro-broadcasting or home-based audio mastering.

Objective Influence on the Signal

Emmanuel Deruty and Damien Tardieu performed a systematic study describing the influence of compressors and brickwall limiters on the musical audio signal. The experiment involved four software limiters: Waves L2, Sonnox Oxford Limiter, Thomas Mundt's Loudmax, Blue Cat's Protector, as well as four software compressors: Waves H-Comp, Sonnox Oxford Dynamics, Sonalksis SV-3157, and URS 1970. The study provides objective data on what limiters and compressors do to the audio signal.

Five signal descriptors were considered: RMS power, EBU3341/R128 integrated loudness, crest factor, EBU3342 LRA, and density of clipped samples. RMS power accounts for the signal's physical level, EBU3341 loudness for the perceived level. The crest factor, which is the difference between the signal's peak and its average power, is on occasions considered as a basis for the measure of micro-dynamics, for instance in the *TT Dynamic Range Meter* plug-in. Finally, EBU3342 LRA has been repeatedly considered as a measure of macro-dynamics or dynamics in the musical sense.

Limiters

The tested limiters had the following influence on the signal:

- increase of RMS power,
- increase of EBU3341 loudness,
- decrease of crest factor,
- decrease of EBU3342 LRA, but only for high amounts of limiting,
- increase of clipped sample density.

In other words, limiters increase both physical and perceptual levels, increase the density of clipped samples, decrease the crest factor and decrease macro-dynamics (LRA) given that the amount of limiting is substantial.

Compressors

As far as the compressors are concerned, the authors performed two processing sessions, using a fast attack (0.5 ms) in one case, and a slow attack (50 ms) in the other. Gain make-up is deactivated, but the resulting file is normalized.

Set with a fast attack, the tested compressors had the following influence on the signal:

- slight increase of RMS power,
- slight increase of EBU3341 loudness,
- decrease of crest factor,
- decrease of EBU3342 LRA,
- slight decrease of clipped sample density.

In other words, fast-attack compressors increase both physical and perceptual levels, but only slightly. They decrease the density of clipped samples, and decrease both crest factor and macro-dynamics.

Set with a slow attack, the tested compressors had the following influence on the signal:

- decrease of RMS power,
- decrease of EBU3341 loudness,
- no influence on crest factor,
- decrease of EBU3342 LRA,
- no influence on clipped sample density.

In other words, slow-attack compressors decrease both physical and perceptual levels, decrease macro-dynamics, but have no influence on crest factor and clipped sample density.

Audio Time Stretching and Pitch Scaling

Pitch Shifting

As opposed to the process of pitch transposition achieved using a simple sample rate conversion, Pitch Shifting is a way to change the pitch of a signal without changing its length. In practical applications, this is usually achieved by changing the length of a sound using one of the below methods and then performing a sample rate conversion to change the pitch.

There exists a certain confusion in terminology, as Pitch Shifting is often also incorrectly named 'Frequency Shifting'. A true Frequency Shift (as obtainable by modulating an analytic signal by a complex exponential) will shift the spectrum of a sound, while Pitch Shifting will dilate it, upholding the harmonic relationship of the sound. Frequency Shifting yields a metallic, inharmonic sound which may well be an interesting special effect but which is a totally inadequate process for changing the pitch of any harmonic sound except a single sine wave.

Time Compression/Expansion

Time Compression/Expansion, also known as "Time Stretching" is the reciprocal process to Pitch Shifting. It leaves the pitch of the signal intact while changing its speed (tempo). This is a useful application when you wish to change the speed of a voiceover without messing with the timbre of the voice.

There are several fairly good methods to do time compression/expansion and pitch shifting but most of them will not perform well on all different kinds of signals and for any desired amount of shift/stretch ratio. Typically, good algorithms allow pitch shifting up to 5 semitones on average or stretching the length by 130%. When time stretching and pitch shifting single instrument recordings you might even be able to achieve a 200% time stretch, or a one-octave pitch shift with no audible loss in quality.

Resampling

The simplest way to change the duration or pitch of a digital audio clip is through sample rate conversion. This is a mathematical operation that effectively rebuilds a continuous waveform from its samples and then samples that waveform again at a different rate. When the new samples are played at the original sampling frequency, the audio clip sounds faster or slower. Unfortunately, the frequencies in the sample are always scaled at the same rate as the speed, transposing its perceived pitch up or down in the process. In other words, slowing down the recording lowers the pitch, speeding it up raises the pitch. This is analogous to speeding up or slowing down an analogue recording, like a phonograph record or tape, creating the Chipmunk effect. Using this method the two effects cannot be separated. A drum track containing no pitched instruments can be moderately sample rate converted for tempo without adverse effects, but a pitched track cannot.

Time Domain

Sola

Rabiner and Schafer in 1978 put forth an alternate solution that works in the time domain: attempt to find the period (or equivalently the fundamental frequency) of a given section of the wave using some pitch detection algorithm (commonly the peak of the signal's autocorrelation, or sometimes cepstral processing), and crossfade one period into another.

This is called time-domain harmonic scaling or the synchronized overlap-add method (SOLA) and performs somewhat faster than the phase vocoder on slower machines but fails when the autocorrelation mis-estimates the period of a signal with complicated harmonics (such as orchestral pieces).

Adobe Audition (formerly Cool Edit Pro) seems to solve this by looking for the period closest to a center period that the user specifies, which should be an integer multiple of the tempo, and between 30 Hz and the lowest bass frequency.

This is much more limited in scope than the phase vocoder based processing, but can be made much less processor intensive, for real-time applications. It provides the most coherent results for single-pitched sounds like voice or musically monophonic instrument recordings.

High-end commercial audio processing packages either combine the two techniques (for example by separating the signal into sinusoid and transient waveforms), or use other techniques based on the wavelet transform, or artificial neural network processing, producing the highest-quality time stretching.

Frame-based Approach

In order to preserve an audio signal's pitch when stretching or compressing its duration, many time-scale modification (TSM) procedures follow a frame-based approach. Given an original discrete-time audio signal, this strategy's first step is to split the signal into short *analysis frames* of fixed length. The analysis frames are spaced by a fixed number of samples, called the *analysis hopsize* $H_a \in \mathbb{N}$. To achieve the actual time-scale modification, the analysis frames are then temporally relocated to have a *synthesis hopsize* $H_s \in \mathbb{N}$. This frame relocation results in a modification of the signal's duration by a *stretching factor* of $\alpha = H_s / H_a$. However, simply superimposing the unmodified analysis frames typically results in undesired artifacts such as phase discontinuities or amplitude fluctuations. To prevent these kinds of artifacts, the analysis frames are adapted to form *synthesis frames*, prior to the reconstruction of the time-scale modified output signal.

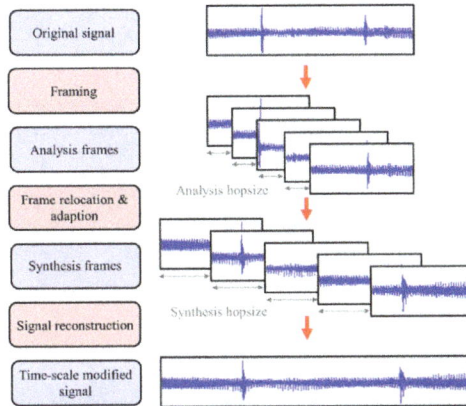

Frame-based approach of many TSM procedures

The strategy of how to derive the synthesis frames from the analysis frames is a key difference among different TSM procedures.

Frequency Domain

Phase Vocoder

One way of stretching the length of a signal without affecting the pitch is to build a phase vocoder after Flanagan, Golden, and Portnoff.

Basic steps:

1. compute the instantaneous frequency/amplitude relationship of the signal using the STFT, which is the discrete Fourier transform of a short, overlapping and smoothly windowed block of samples;

2. apply some processing to the Fourier transform magnitudes and phases (like resampling the FFT blocks); and

3. perform an inverse STFT by taking the inverse Fourier transform on each chunk and adding the resulting waveform chunks, also called overlap and add (OLA).

The phase vocoder handles sinusoid components well, but early implementations introduced considerable smearing on transient ("beat") waveforms at all non-integer compression/expansion rates, which renders the results phasey and diffuse. Recent improvements allow better quality results at all compression/expansion ratios but a residual smearing effect still remains.

The phase vocoder technique can also be used to perform pitch shifting, chorusing, timbre manipulation, harmonizing, and other unusual modifications, all of which can be changed as a function of time.

Sinusoidal Spectral Modeling

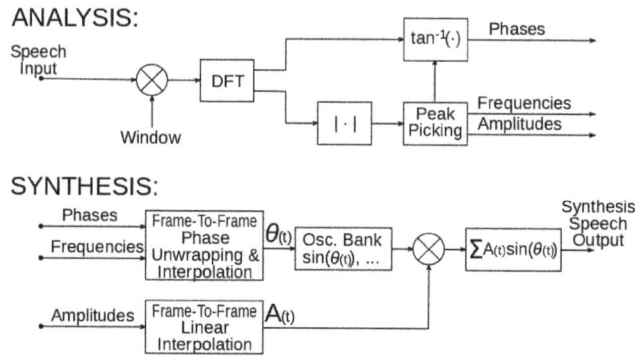

Sinusoidal analysis/synthesis system

Another method for time stretching relies on a spectral model of the signal. In this method, peaks are identified in frames using the STFT of the signal, and sinusoidal "tracks" are created by connecting peaks in adjacent frames. The tracks are then re-synthesized at a new time scale. This method can yield good results on both polyphonic and percussive material, especially when the signal is separated into subbands. However, this method is more computationally demanding than other methods.

Speed Hearing and Speed Talking

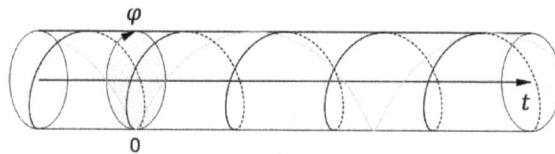

Modelling a monophonic sound as observation along a helix
of a function with a cylinder domain

For the specific case of speech, time stretching can be performed using PSOLA.

While one might expect speeding up to reduce comprehension, Herb Friedman says that "Experiments have shown that the brain works most efficiently if the information rate through the ears--via speech--is the "average" reading rate, which is about 200-300 wpm (words per minute), yet the average rate of speech is in the neighborhood of 100-150 wpm."

Speeding up audio is seen as the equivalent of "speed reading".

Time stretching is often used to adjust Radio commercials and the audio of Television advertisements to fit exactly into the 30 or 60 seconds available.

Pitch Scaling

Pitch shifting (Frequency scaling) is provided on Eventide Harmonizer

Frequency shifting provided by Bode Frequency Shifter *does not keep* frequency ratio and harmony.

These techniques can also be used to transpose an audio sample while holding speed or duration constant. This may be accomplished by time stretching and then resampling back to the original length. Alternatively, the frequency of the sinusoids in a sinusoidal model may be altered directly, and the signal reconstructed at the appropriate time scale.

Transposing can be called *frequency scaling* or *pitch shifting*, depending on perspective.

For example, one could move the pitch of every note up by a perfect fifth, keeping the tempo the same. One can view this transposition as "pitch shifting", "shifting" each note up 7 keys on a piano keyboard, or adding a fixed amount on the Mel scale, or adding a fixed amount in linear pitch space. One can view the same transposition as "frequency scaling", "scaling" (multiplying) the frequency of every note by 3/2.

Musical transposition preserves the ratios of the harmonic frequencies that determine the sound's timbre, unlike the *frequency shift* performed by amplitude modulation, which adds a fixed frequency offset to the frequency of every note. (In theory one could perform a literal *pitch scaling* in which the musical pitch space location is scaled [a higher note would be shifted at a greater interval in linear pitch space than a lower note], but that is highly unusual, and not musical).

Time domain processing works much better here, as smearing is less noticeable, but scaling vocal samples distorts the formants into a sort of Alvin and the Chipmunks-like effect, which may be desirable or undesirable. A process that preserves the formants and character of a voice involves analyzing the signal with a channel vocoder or LPC vocoder plus any of several pitch detection algorithms and then resynthesizing it at a different fundamental frequency.

A detailed description of older analog recording techniques for pitch shifting can be found within the Alvin and the Chipmunks entry.

Audio Mixing

Audio mixing is the process of taking recorded tracks and blending them together. Tracks are blended using various processes such as EQ, Compression and Reverb.

The goal of mixing is to bring out the best in your multi-track recording by adjusting levels, panning, and time-based effects (chorus, reverb, delay). The aim is to sculpt your arrangement to make sense of all your tracks in relation to each other.

A multitrack recording is anything with more than one individual track (also referred to as stems). There's no right or wrong number of tracks. You just can't have zero. The final output of a multitrack recording is also known as the mixdown. The mixdown is the final step before mastering.

It doesn't matter if you're recording tracks with microphones and pre-amps, or using pre-recorded samples, learning how to mix for yourself is insanely important. Taking control of your artistic and creative vision will take your music to the next level.

Audio mixing is practiced for music, film, television and live sound. The process is generally carried out by a mixing engineer operating a mixing console or digital audio workstation.

Recorded Music

Before the introduction of multitrack recording, all the sounds and effects that were to be part of a recording were mixed together at one time during a live performance. If the sound blend was not satisfactory, or if one musician made a mistake, the selection had to be performed over until the desired balance and performance was obtained. However, with the introduction of multitrack recording, the production phase of a modern recording has radically changed into one that generally involves three stages: recording, overdubbing, and mixdown.

Film And Television

Audio console in a cable news control room.

Audio mixing for film and television is a process during the post-production stage of a moving image program by which a multitude of recorded sounds are combined. In the process, the source's signal level, frequency content, dynamics and panoramic position are commonly manipulated and effects added.

The process takes place on a mix stage, typically in a studio or theater, once the picture elements are edited into a final version. Normally the engineer will mix four main audio elements: speech (dialogue, ADR, voice-overs, etc.), ambience (or atmosphere), sound effects, and music.

Live Sound

Live sound mixing is the process of electrically blending together multiple sound sources at a live event using a mixing console. Sounds used include those from instruments, voices, and pre-recorded material. Individual sources may be equalised and routed to effect processors to ultimately be amplified and reproduced via loudspeakers. The live sound engineer balances the various audio sources in a way that best suits the needs of the event.

Audio Post Production

Audio Post Production is the process of creating the soundtrack for moving images. Ever since the once silent movies developed a prerecorded track, filmmakers have been looking to control and improve the quality of the sound of their visions. As soon as moviemakers realized there was a way to control and enhance the sound of their pictures, Audio Post was born and has been a fact of life ever since. In television, audio was originally "live," like the visual program it was part of. As TV evolved and grew to include "videotaped" and "filmed" programming, its need for audio post increased. Nowadays, it would be difficult to find any feature film or television show (or video game) that hasn't been through audio post.

Processes of Audio Post

Audio post usually consists of several processes. Each different project may need some or all of these processes in order to be complete. The processes are:

- Production Dialogue Editing - In order for the production audio recorded on the set or on location to be properly mixed, a Dialogue Editor needs to prepare it. This means locating the takes used by the Picture Editor from the recorded production audio, checking sync (so the audio works with the picture properly), and eliminating extraneous noise so the Dialogue Mixer has clean dialogue to use during the mix.

- ADR [Automated Dialogue Replacement] - In cases where the production audio is too noisy or otherwise unusable (bad line reading, airplane fly-by, etc.), or where the filmmakers want to add voice over narration or simply add dialogue that was never recorded, the line will be programmed or "cued" for "looping" or ADR. This process takes place on the ADR Stage, a specialized recording studio where the actor can record while watching the edited picture, matching the sync of the original line or fitting the new lines with the actions.

- After a loop lines have been recorded, the ADR Editor will check the sync carefully, modifying the take if necessary to precisely match it to the picture, and prepare it for the Mixing Stage.

- Sound Effects Design and Editing - Ever wonder how they made the sound of Darth Vader's helmet breath, or the roar of Jurassic dinosaurs, or those great explosions that seem to get bigger every year? Sound Effects Editors and Sound Designers are how. They are the craftspeople who add the computer beeps, gunshots, laser blasts, massive explosions; and more subtle sounds like background ambiences such as air, rivers, birds, and city traffic. Sound Designers use a variety of technologies from bleeding edge to tried & true to create unique sound effects that have never been heard before, or to artistically create specific "mood" sounds to complement the filmmakers' vision of the visuals. Sound Effects Editors put those sounds in sync with the picture as well as selecting from libraries of hundreds of thousands of prerecorded sounds; and organize them so the FX Mixers can "PreDubb" those sounds efficiently.

- Foley - Taking its name from Jack Foley, the Hollywood sound editor regarded as the "father" of these effects, Foley effects are sounds that are created by recording (usually) everyday movement while watching the edited picture. Different from the environmental backgrounds ("BGs") and hard effects (FX), Foley effects are sounds like footsteps, object handling, the rustling of clothing, etc. The people involved in this process are the Foley Walkers or Artists who perform those sounds and the Foley Mixer who records them. After the Foley Effects are "shot," the Foley Editor will use his/her craft to polish those sounds to ensure that they are exactly in sync with the final picture.

- Music Composition - Music for motion pictures falls into two general categories: Score and Source. The Composer is the individual hired to prepare the dramatic underscore. Source music is what we hear coming from an on screen or off screen device like stereos, televisions, ice cream trucks, and so on. Source music may be original or licensed from a number of libraries that specialize in the creation of "generic" music. Songs (music with vocals) may occupy either function, depending on the dramatic intent of the director. For "Pulp Fiction" for example, Director Quentin Tarantino hired a Music Supervisor (Karyn Rachtman) to "score" the picture using period music of the 1970's almost exclusively. Most contemporary films use a combination of score and source music.

- Music Editing - The Music Editor assists the Composer in the preparation of the dramatic underscore. Frequently working also with the Music Supervisor, the Music Editor will take timings for the Composer during a spotting session in order to notate the specific locations in the film where underscore or source music will punctuate the narrative. Once the underscore is recorded and the source music gathered, the Music Editor would be the person who edits or supervises the final synchronization of all music elements prior to the mix.

- Mixing (also called Dubbing) - The Mixers have the responsibility of balancing the various elements, i.e., the Dialogue & ADR, Music, Sound Effects, and Foley Effects, in the final mix. The Dialogue Mixer, (also called the Lead Mixer or Gaffing Mixer) commands the mixing stage; his/her partners in the mix traditionally were the Effects Mixer and the Music Mixer. As of now, the Lead Mixer commonly does the Music mixing as well, reducing the traditional mixing team by a third. On huge pictures with tight deadlines, it is possible that several teams of mixers are working simultaneously on numerous stages in order to complete the mix by the release date.

Audio Mastering

Mastering is the final step of audio post-production. The purpose of mastering is to balance sonic elements of a stereo mix and optimize playback across all systems and media formats. Traditionally, mastering is done using tools like equalization, compression, limiting and stereo enhancement.

Think of mastering as the glue, varnish and polish that optimizes playback quality on all devices — from tiny iPhone speakers to massive dance club sound systems. Mastering bridges the gap between artist and consumer. The term itself comes from the idea of a master copy. All copies or duplications of the audio come from the master. These copies can be distributed on multiple formats like vinyl, CD's or Tape, and streaming services like Spotify, iTunes and SoundCloud. Additionally, mastering allows for restoration of hisses, clicks or small mistakes missed in the final mix. Mastering also ensures uniformity and consistency of sound between multiple tracks on an album. Ultimately, what mastering does is create a clean and cohesive feeling across all your audio.

The goal of mastering is to ensure that your audio will sound the best it can on all platforms. Music has never been consumed on more formats and devices than today. Even if you are recording and mixing in a million dollar studio, or recording in less than ideal conditions, you still need the final quality check of mastering. This ensures that your sound will be heard the way you intended it to be.

A good mastering job makes an album consistent and balanced across all tracks. Without mastering, individual tracks can sound disjointed in relation to each other.

Digital Technology

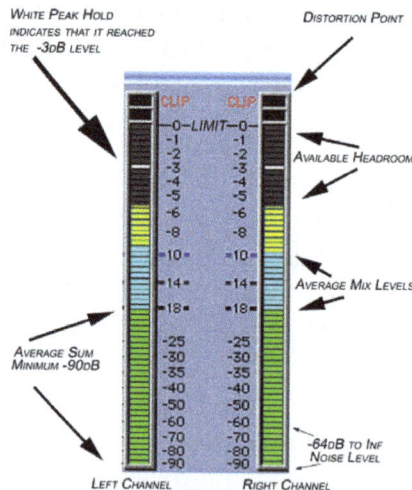

Optimum Digital Levels with respect to the Full Digital Scale (dBFSD)

In the 1990s, electro-mechanical processes were largely superseded by digital technology, with digital recordings stored on hard disk drives or digital tape and mastered to CD. The digital audio workstation (DAW) became common in many mastering facilities, allowing the off-line manipulation of recorded audio via a graphical user interface (GUI). Although many digital processing tools are common during mastering, it is also very common to use analog media and processing equipment for the mastering stage. Just as in other areas of audio, the benefits and drawbacks of digital technology compared to analog technology are still a matter for debate. However, in the field of audio mastering, the debate is usually over the use of digital versus analog signal processing rather than the use of digital technology for storage of audio.

Digital systems have higher performance and allow mixing to be performed at lower maximum levels. With peaks between -3 and -9 dBFS on a mix, the mastering engineer has enough headroom to process and produce a final master. It is important to allow enough headroom for the mastering engineer's work. Reduction of headroom by the mix or mastering engineer has resulted in a loudness war in commercial recordings.

Mastering has a special significance for house, techno and other styles of electronic dance music:

> The product of mastering, regardless of the type of format, is re-actualised into a performance in the hands of the DJ. In this regard, the shift from vinyl records to digital formats in many DJ scenes and the appearance of new digital formats for DJs have encouraged new practices also among mastering engineers, whose targets are not only music listeners but, specifically, music performers as well.

Process

A common mastering processor for dynamic range compression

The source material, ideally at the original resolution, is processed using equalization, compression, limiting, noise reduction and other processes. More tasks, such as editing, pre-gapping, leveling, fading in and out, noise reduction and other signal restoration and enhancement processes can be applied as part of the mastering stage. This step prepares the music for either digital or analog, e.g. vinyl, replication. The source material is put in the proper order, commonly referred to as assembly (or 'track') sequencing.

If the material is destined for vinyl release, additional processing, such as dynamic range reduction or frequency dependent stereo–to–mono fold-down and equalization, may be applied to compensate for the limitations of that medium. Finally, for compact disc release, Start of Track, End of Track, and Indexes are defined for disc navigation. Subsequently, it is rendered either to a physical medium, such as a CD-R or DVD-R, or to computer files, such as a Disc Description Protocol (DDP) file set or an ISO file. The specific medium varies, depending on the intended release format of the final product. For digital audio releases, there is more than one possible master medium, chosen based on replication factory requirements or record label security concerns. Regardless of what delivery method is chosen, the replicator will transfer the audio to a glass master that will generate metal stampers for replication.

The process of audio mastering varies depending on the specific needs of the audio to be processed. Mastering engineers need to examine the types of input media, the expectations of the source producer or recipient, the limitations of the end medium and process the subject accordingly. General rules of thumb can rarely be applied.

Steps of the process typically include the following:

1. Transferring the recorded audio tracks into the Digital Audio Workstation (DAW).

2. Sequence the separate songs or tracks as they will appear on the final release.

3. Adjust the length of the silence between songs.

4. Process or "sweeten" audio to maximize the sound quality for the intended medium (e.g. applying specific EQ for vinyl).

5. Transfer the audio to the final master format (CD-ROM, half-inch reel tape, PCM 1630 U-matic tape, etc.).

Examples of possible actions taken during mastering:

1. Editing minor flaws.

2. Applying noise reduction to eliminate clicks, dropouts, hum and hiss.

3. Adjusting stereo width.

4. Adding ambience.

5. Equalize audio across tracks for the purpose of optimized frequency distribution.

6. Adjust volume.

7. Dynamic range compression or expansion.

8. Peak limit.

9. Dither.

To finish mastering a CD, track markers must be inserted, along with International Standard Recording Code (ISRC) and other information necessary to replicate a CD. Vinyl LP and cassettes have their own pre-duplication requirements for a finished master.

So What Does Mastering Do?

Mastering is a complex process. Here are techniques involved:

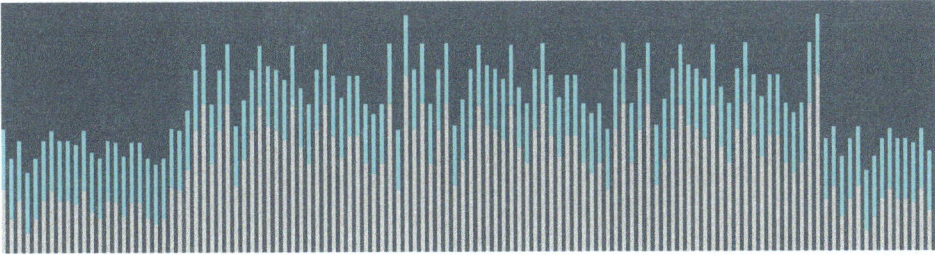

- Audio Restoration

  This step fixes any hiccups in the original mix like unwanted clicks, pops or hisses. It also helps to fix small mistakes that stand out when un-mastered audio is amplified.

- Stereo Enhancement

  Stereo enhancement deals with the spatial balance (left to right) of your audio. Done right, stereo enhancement widens your mix, helping it sound bigger. It can also help tighten your center image by focusing the low-end.

- EQ

  EQing corrects any spectral imbalances and enhances elements that need to stand out. An ideal master is well-balanced and proportional. This means no specific frequency range is left sticking out. A balanced piece of audio will sound good on any playback system.

- Compression

  Compression corrects and enhances the dynamic range of your mix and keeps louder signals in check while bringing up quieter parts. This process gives the overall audio a better uniformity and feel. Compression helps glue together parts that might not be as cohesive as they could be.

- Loudness

  The last process in the mastering chain is usually a special type of compressor called a limiter. Limiters set appropriate overall loudness and creates a peak ceiling. Limiting makes the track competitively loud without allowing any clipping that can lead to distortion.

- Bit Depth Reduction & Sample Rate Conversion

  Sample rate conversion or dither is dependent on the final output medium. For example, if you are planning to release on CD you will have to convert to 44.1kHz 16 bit and therefore, you may have to convert and dither your file to get to the standard of format.

- Sequencing & Spacing

  Sequencing and spacing is one of the final steps in mastering. On an album or EP this process puts your audio in order. Spacing refers to how much silence (space gaps) you put between each track.

Sound Design

Sound design is the art and practice of creating, collecting, adapting, and producing audio elements. Like any other type of design, sound design is a process that involves turning an idea into an outcome. What makes sound design different are the tools used in that process. Whereas a graphic designer's toolkit includes typography, images, and colors, the essential tools for sound designers are music, voices, and sounds.

In certain industries, like film, radio, and video games, it is taken for granted that sound design is integral to developing the narrative. Sound design is strategically employed to demarcate time and space, evoke emotion, create mood, and intensify action.

Thanks to the burgeoning field of sound studies, today, the importance of sound in almost every aspect and realm of human experience is being increasingly understood. With its power to affect setting, provoke feelings, and create memories, sound has been elevated from an afterthought to a cultural artifact worthy of study. By turning a critical lens to sound, we can better understand what it does – how it affects us and compels us to act. From that understanding, we can begin to imagine how we can better design sonic experiences both for today and in the future.

Applications

Film

In motion picture production, a *Sound Editor/Designer* is a member of a film crew responsible for the entirety or some specific parts of a film's sound track. In the American film industry, the title *Sound Designer* is not controlled by any professional organization, unlike titles such as Director or Screenwriter.

The terms *sound design* and *sound designer* began to be used in the motion picture industry in 1979. At that time, The title of *Sound Designer* was first granted to Walter

Murch by Francis Ford Coppola in recognition for Murch's contributions to the film Apocalypse Now. The original meaning of the title *Sound Designer*, as established by Coppola and Murch, was "an individual ultimately responsible for all aspects of a film's audio track, from the dialogue and sound effects recording to the re-recording (mix) of the final track". The term *sound designer* has replaced monikers like *supervising sound editor* or *re-recording mixer* for what was essentially the same position: the head designer of the final sound track. Editors and mixers like Murray Spivack (*King Kong*), George Groves (*The Jazz Singer*), James G. Stewart (*Citizen Kane*), and Carl Faulkner (*Journey to the Center of the Earth*) served in this capacity during Hollywood's studio era, and are generally considered to be sound designers by a different name.

The advantage of calling oneself a sound designer beginning in later decades was two-fold. It strategically allowed for a single person to work as both an editor and mixer on a film without running into issues pertaining to the jurisdictions of editors and mixers, as outlined by their respective unions. Additionally, it was a rhetorical move that legitimated the field of post-production sound at a time when studios were down-sizing their sound departments, and when producers were routinely skimping on budgets and salaries for sound editors and mixers. In so doing, it allowed those who called themselves sound designers to compete for contract work and to negotiate higher salaries. The position of Sound Designer therefore emerged in a manner similar to that of Production Designer, which was created in the 1930s when William Cameron Menzies made revolutionary contributions to the craft of art direction in the making of *Gone with the Wind*.

The audio production team is a principal member of the production staff, with creative output comparable to that of the film editor and director of photography. Several factors have led to the promotion of audio production to this level, when previously it was considered subordinate to other parts of film:

- Cinema sound systems became capable of high-fidelity reproduction, particularly after the adoption of Dolby Stereo. These systems were originally devised as gimmicks to increase theater attendance, but their widespread implementation created a content vacuum that had to be filled by competent professionals. Before stereo soundtracks, film sound was of such low fidelity that only the dialogue and occasional sound effects were practical. The greater dynamic range of the new systems, coupled with the ability to produce sounds at the sides or behind the audience, provided the audio production team new opportunities for creative expression in film sound.

- Some directors were interested in realizing the new potentials of the medium. A new generation of filmmakers, the so-called "Easy Riders and Raging Bulls"—Martin Scorsese, Steven Spielberg, George Lucas, and others—were aware of the creative potential of sound and wanted to use it.

- Filmmakers were inspired by the popular music of the era. Concept albums of groups such as Pink Floyd and The Beatles suggested new modes of storytelling and creative techniques that could be adapted to motion pictures.

- New filmmakers made their early films outside the Hollywood establishment, away from the influence of film labor unions and the then rapidly dissipating studio system.

The contemporary title of *sound designer* can be compared with the more traditional title of *supervising sound editor*; many sound designers use both titles interchangeably. The role of supervising sound editor, or sound supervisor, developed in parallel with the role of sound designer. The demand for more sophisticated soundtracks was felt both inside and outside Hollywood, and the supervising sound editor became the head of the large sound department, with a staff of dozens of sound editors, that was required to realize a complete sound job with a fast turnaround.

Theatre

Sound design, as a distinct discipline, is one of the youngest fields in stagecraft, second only to the use of projection and other multimedia displays, although the ideas and techniques of sound design have been around almost since theatre started. Dan Dugan, working with three stereo tape decks routed to ten loudspeaker zones during the 1968–69 season of American Conservatory Theater (ACT) in San Francisco, was the first person to be called a sound designer.

Modern audio technology has enabled theatre sound designers to produce flexible, complex, and inexpensive designs that can be easily integrated into live performance. The influence of film and television on playwriting is seeing plays being written increasingly with shorter scenes, which is difficult to achieve with scenery but easily conveyed with sound. The development of film sound design is giving writers and directors higher expectations and knowledge of sound design. Consequently, theatre sound design is widespread and accomplished sound designers commonly establish long-term collaborations with directors.

Musicals

Sound design for musicals often focuses on the design and implementation of a sound reinforcement system that will fulfill the needs of the production. If a sound system is already installed in the performance venue, it is the sound designer's job to tune the system for the best use for a particular production. Sound system tuning employs various methods including equalization, delay, volume, speaker and microphone placement, and in some cases, the addition of new equipment. In conjunction with the director and musical director, if any, the sound reinforcement designer determines the use and placement of microphones for actors and musicians. The sound reinforcement designer ensures that the performance can be heard and understood by everyone in the audience, regardless of the shape, size or acoustics of the venue, and that performers can hear everything needed to enable them to do their jobs. While sound design for a musical largely focuses on the artistic merits of sound reinforcement, many musicals, such as *Into the Woods* also require significant sound scores. Sound Reinforcement Design was recognized by the American Theatre Wing's Tony Awards with the Tony Award for Best Sound Design of a Musical until the 2014-15 season, later reinstating in the 2017-18 season.

Plays

Sound design for plays often involves the selection of music and sounds (sound score) for a production based on intimate familiarity with the play, and the design, installation, calibration and utilization of the sound system that reproduces the sound score. The sound designer for a play and the production's director work together to decide the themes and emotions to be explored. Based on this, the sound designer for plays, in collaboration with the director and possibly the composer, decides upon the sounds that will be used to create the desired moods. In some productions, the sound designer might also be hired to compose music for the play. The sound designer and the director usually work together to "spot" the cues in the play (i.e., decide when and where sound will be used in the play). Some productions might use music only during scene changes, whilst others might use sound effects. Likewise a scene might be underscored with music, sound effects or abstract sounds that exist somewhere between the two. Some sound designers are accomplished composers, writing and producing music for productions as well as designing sound. Many sound designs for plays also require significant sound reinforcement. Sound Design for plays was recognized by the American Theatre Wing's Tony Awards with the Tony Award for Best Sound Design of a Play until the 2014-15 season, later reinstating the award in the 2017-18 season.

Music

In contemporary music business, especially in the production of rock music, ambient music, progressive rock, and similar genres, the record producer and recording engineer play important roles in the creation of the overall sound (or soundscape) of a recording, and less often, of a live performance. A record producer is responsible for extracting the best performance possible from the musicians and for making both musical and technical decisions about the instrumental timbres, arrangements, etc. On some, particularly more electronic music projects, artists and producers in more conventional genres have sometimes sourced additional help from artists often credited as "sound designers", to contribute specific auditory effects, ambiences etc. to the production. These people are usually more versed in e.g. electronic music composition and synthesizers than the other musicians on board.

In application of electroacoustic techniques (e.g. binaural sound) and sound synthesis for contemporary music or film music, a sound designer (often also an electronic musician) sometimes refers to an artist who works alongside a composer to realize the more electronic aspects of a musical production. This is because sometimes there exists a difference in interests between artists calling themselves composers and artists calling themselves electronic musicians or sound designers. The latter being sometimes more experienced as well as interested in electronic music techniques, such as sequencing and synthesizers, but the former often wanting to use elements of electronic music in compositions but often being more experienced in writing music in a variety of genres. Since electronic music itself is quite broad in techniques and often separate from techniques applied in other genres, this kind of collaboration can be seen as fairly natural as well as beneficial.

Notable examples of (recognized) sound design in music are the contributions of Michael Brook to the U2 album *The Joshua Tree*, George Massenburg to the Jennifer Warnes album *Famous Blue Raincoat*, Chris Thomas to the Pink Floyd album *The Dark Side of the Moon*, and Brian Eno to the Paul Simon album *Surprise*.

In 1974, Suzanne Ciani started her own production company, Ciani/Musica. Inc., which became the #1 sound design music house in New York.

Computer Applications and Other Applications

Sound is widely used in a variety of human-computer interfaces, in computer games and video games. Sound production for computer applications adds few extra concerns and requirements to the sound. Some of them are: non-repetitiveness, interactivity/dynamicity and low memory and CPU use. For example, the most computational resources are usually devoted to graphics, which in turn means that the sound playback is often limited to some upper memory consumption as well as CPU use limits. These have to be added to the concerns of audio production, since it is often not enough to merely "create good sound", but the sound also has to fit to the given computational limits, while still sounding good, which may require e.g. the use of audio compression or voice allocating systems. However, sound for computer applications also adds some new creative aspects to music and sound, because computer sound (especially in games) often involves the sound having or desired to be *interactive*. Adding interactivity can involve using a variety of playback systems or logic, using tools that allow the production of interactive sound (e.g. Max/MSP, Wwise) and it also deviates from being just an art form to also requiring software or electrical engineering, since implementing interactivity of sound also requires engineering of the systems that play the sound or e.g. process user input. Therefore, in interactive applications, a sound designer often collaborates with an engineer (e.g. a sound programmer) who's concerned with designing the playback systems and their efficiency.

Examples

The eerie scream of an Alien, or the sound of the weapons going off in a video game. A sound designer is one who creates all the stunning sound SFX. Sound design requires both technological and artistic skills. A sound designer needs to develop excellent knowledge and techniques in recording, mixing, and special effects in order to create unique and interesting sounds. There are a host of modern Software and hardware synths which are used to create new and unheard SFX and sounds.

In conclusion, good sound design has the ability to create sonic experiences that are not just aesthetically pleasing, but also purposeful. For brands, using strategic sound design can extend branded product and service experiences beyond the visual realm. Because sound is strongly linked to emotion, a product that emits appealing sounds is more likely to strike a chord in the hearts of users. Since sound has the power to set the tone of how customers feel when they're in a store, it can subtly affect purchasing behavior. And as memories can be both created and elicited from auditory experiences, skillful sound design contributes to the consumer's lasting impression of a brand. With such compelling reasons, it's not surprising that many brands are now considering how the sound of their brand can be designed to create a more holistic and lasting brand experience.

Sound Effect

Sound effect, any artificial reproduction of sound or sounds intended to accompany action and supply realism in the theatre, radio, television, and motion pictures. Sound effects have traditionally been of great importance in the theatre, where many effects, too vast in scope, too dangerous, or simply too expensive to be presented on stage, must be represented as taking place behind the scenes. An offstage battle, for instance, can be simulated by such sounds as trumpet blasts, shouts, shots, clashing weapons, and horses' hooves. Certain dangerous effects, such as explosions, crashes, and the smashing of wood or glass, must also take place offstage. Sound effects must often be coordinated with actions on stage; when the hero pretends to punch the villain on the jaw, a sound technician backstage must provide a realistic "smack!"

Many ingenious methods have been devised for the faithful reproduction of various sounds; wind sounds, from a breeze to a hurricane, can be simulated when a piece of canvas is rubbed by wooden slats mounted on a revolving cylinder; thunder is imitated by shaking a large sheet of metal; rain sounds are produced by rattling dried peas in a wooden box; horses' hooves can be imitated by clattering coconut shells or suction cups against a hard surface; gunshots can be produced by slapping boards together or by firing blank cartridges.

Today most sound effects are recorded on records or tapes, which provide greater realism and allow for the production of an almost limitless range of effects with no need of bulky sound-producing devices.

Film

In the context of motion pictures and television, *sound effects* refers to an entire hierarchy of sound elements, whose production encompasses many different disciplines, including:

- *Hard sound effects* are common sounds that appear on screen, such as door alarms, weapons firing, and cars driving by.

- *Background* (or *BG*) *sound effects* are sounds that do not explicitly synchronize with the picture, but indicate setting to the audience, such as forest sounds, the buzzing of fluorescent lights, and car interiors. The sound of people talking in the background is also considered a "BG," but only if the speaker is unintelligible and the language is unrecognizable (this is known as walla). These background noises are also called *ambience* or *atmos* ("atmosphere").

- *Foley sound effects* are sounds that synchronize on screen, and require the expertise of a foley artist to record properly. Footsteps, the movement of hand props (e.g., a tea cup and saucer), and the rustling of cloth are common foley units.

- *Design sound effects* are sounds that do not normally occur in nature, or are impossible to record in nature. These sounds are used to suggest futuristic technology in a science fiction film, or are used in a musical fashion to create an emotional mood.

Each of these sound effect categories is specialized, with sound editors known as specialists in an area of sound effects (e.g. a "Car cutter" or "Guns cutter").

Foley is another method of adding sound effects. Foley is more of a technique for creating sound effects than a type of sound effect, but it is often used for creating the incidental real world sounds that are very specific to what is going on onscreen, such as footsteps. With this technique the action onscreen is essentially recreated to try to match it as closely as possible. If done correctly it is very hard for audiences to tell what sounds were added and what sounds were originally recorded (location sound).

In the early days of film and radio, foley artists would add sounds in realtime or pre-recorded sound effects would be played back from analogue discs in realtime (while watching the picture). Today, with effects held in digital format, it is easy to create any required sequence to be played in any desired timeline.

In the days of silent film, sound effects were added by the operator of a theater organ or photoplayer, both of which also supplied the soundtrack of the film. Theater organ sound effects are usually electric or electro-pneumatic, and activated by a button pressed with the hand or foot. Photoplayer operators activate sound effects either by flipping switches on the machine or pulling "cow-tail" pull-strings, which hang above. Sounds like bells and drums are made mechanically, sirens and horns electronically.

Due to its smaller size, a photoplayer usually has less special effects than a theater organ, or less complex ones.

Video Games

The principles involved with modern video game sound effects (since the introduction of sample playback) are essentially the same as those of motion pictures. Typically a game project requires two jobs to be completed: sounds must be recorded or selected from a library and a sound engine must be programmed so that those sounds can be incorporated into the game's interactive environment.

In earlier computers and video game systems, sound effects were typically produced using sound synthesis. In modern systems, the increases in storage capacity and playback quality has allowed sampled sound to be used. The modern systems also frequently utilize positional audio, often with hardware acceleration, and real-time audio post-processing, which can also be tied to the 3D graphics development. Based on the internal state of the game, multiple different calculations can be made. This will allow for, for example, realistic sound dampening, echoes and doppler effect.

Historically the simplicity of game environments reduced the required number of sounds needed, and thus only one or two people were directly responsible for the sound recording and design. As the video game business has grown and computer sound reproduction quality has increased, however, the team of sound designers dedicated to game projects has likewise grown and the demands placed on them may now approach those of mid-budget motion pictures.

Music

Some pieces of music use sound effects that are made by a musical instrument or by other means. An early example is the 18th century Toy Symphony. Richard Wagner in the opera *Das Rheingold* (1869) lets a choir of anvils introduce the scene of the dwarfs who have to work in the mines, similar to the introduction of the dwarfs in the 1937 Disney movie *Snow White*. Klaus Doldingers soundtrack for the 1981 movie *Das Boot* includes a title score with a sonar sound to reflect the U-boat setting. John Barry integrated into the title song of *Moonraker* (1979) a sound representing the beep of a Sputnik like satellite.

Recording

The most realistic sound effects may originate from original sources; the closest sound to machine-gun fire could be an original recording of actual machine guns.

Despite this, real life and actual practice do not always coincide with theory. When recordings of real life do not sound realistic on playback, Foley and f/x are used to create more convincing sounds. For example, the realistic sound of bacon frying can be the crumpling of cellophane, while rain may be recorded as salt falling on a piece of tinfoil.

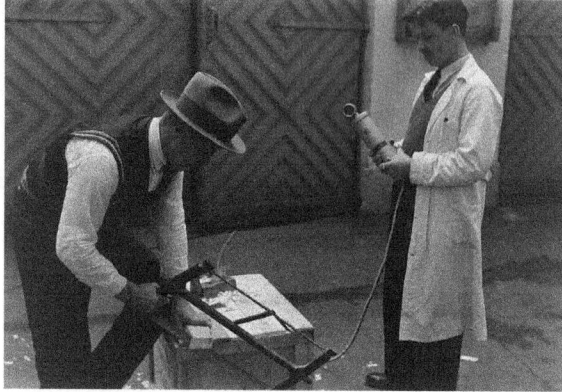

A man recording the sound of a saw

Less realistic sound effects are digitally synthesized or sampled and sequenced (the same recording played repeatedly using a sequencer). When the producer or content creator demands high-fidelity sound effects, the sound editor usually must augment his available library with new sound effects recorded in the field.

When the required sound effect is of a small subject, such as scissors cutting, cloth ripping, or footsteps, the sound effect is best recorded in a studio, under controlled conditions. Such small sounds are often delegated to a foley artist and foley editor. Many sound effects cannot be recorded in a studio, such as explosions, gunfire, and automobile or aircraft maneuvers. These effects must be recorded by a sound effects editor or a professional sound effects recordist.

When such "big" sounds are required, the recordist will begin contacting professionals or technicians in the same way a producer may arrange a crew; if the recordist needs an explosion, he may contact a demolition company to see if any buildings are scheduled to be destroyed with explosives in the near future. If the recordist requires a volley of cannon fire, he may contact historical re-enactors or gun enthusiasts.

Depending on the effect, recordists may use several DAT, hard disk, or Nagra recorders and a large number of microphones. During a cannon- and musket-fire recording session for the 2003 film *The Alamo*, conducted by Jon Johnson and Charles Maynes, two to three DAT machines were used. One machine was stationed near the cannon itself, so it could record the actual firing. Another was stationed several hundred yards away, below the trajectory of the ball, to record the sound of the cannonball passing by. When the crew recorded musket-fire, a set of microphones were arrayed close to the target (in this case a swine carcass) to record the musket-ball impacts.

A counter-example is the common technique for recording an automobile. For recording "Onboard" car sounds (which include the car interiors), a three-microphone technique is common. Two microphones record the engine directly: one is taped to the underside of the hood, near the engine block. The second microphone is covered in a wind screen and tightly attached to the rear bumper, within an inch or so of the tail pipe. The third microphone, which is often a stereo microphone, is stationed inside the car to get the car interior.

Having all of these tracks at once gives a sound designer or audio engineer a great deal of control over how he wants the car to sound. In order to make the car more ominous or low, he can mix in more of the tailpipe recording; if he wants the car to sound like it is running full throttle, he can mix in more of the engine recording and reduce the interior perspective. In cartoons, a pencil being dragged down a washboard may be used to simulate the sound of a sputtering engine. What is considered today to be the first recorded sound effect was of Big Ben striking 10:30, 10:45, and 11:00. It was recorded on a brown wax cylinder by technicians at Edison House in London on July 16, 1890. This recording is currently in the public domain.

Processing Effects

As the car example demonstrates, the ability to make multiple simultaneous recordings of the same subject—through the use of several DAT or multitrack recorders—has made sound recording into a sophisticated craft. The sound effect can be shaped by the sound editor or sound designer, not just for realism, but for emotional effect.

Once the sound effects are recorded or captured, they are usually loaded into a computer integrated with an audio non-linear editing system. This allows a sound editor or sound designer to heavily manipulate a sound to meet his or her needs.

The most common sound design tool is the use of layering to create a new, interesting sound out of two or three old, average sounds. For example, the sound of a bullet impact into a pig carcass may be mixed with the sound of a melon being gouged to add to the "stickiness" or "gore" of the effect. If the effect is featured in a close-up, the designer may also add an "impact sweetener" from his or her library. The sweetener may simply be the sound of a hammer pounding hardwood, equalized so that only the low-end can be heard. The low end gives the three sounds together added weight, so that the audience actually "feels" the weight of the bullet hit the victim.

If the victim is the villain, and his death is climactic, the sound designer may add reverb to the impact, in order to enhance the dramatic beat. And then, as the victim falls over in slow motion, the sound editor may add the sound of a broom whooshing by a microphone, pitch-shifted down and time-expanded to further emphasize the death. If the film is science-fiction, the designer may phaser the "whoosh" to give it a more sci-fi feel.

Aesthetics

When creating sound effects for films, sound recordists and editors do not generally concern themselves with the verisimilitude or accuracy of the sounds they present. The sound of a bullet entering a person from a close distance may sound nothing like the sound designed in the above example, but since very few people are aware of how such a thing actually sounds, the job of designing the effect is mainly an issue of creating a conjectural sound which feeds the audience's expectations while still suspending disbelief.

In the previous example, the phased 'whoosh' of the victim's fall has no analogue in real life experience, but it is emotionally immediate. If a sound editor uses such sounds in the context of emotional climax or a character's subjective experience, they can add to the drama of a situation in a way visuals simply cannot. If a visual effects artist were to do something similar to the 'whooshing fall' example, it would probably look ridiculous or at least excessively melodramatic.

The "Conjectural Sound" principle applies even to happenstance sounds, such as tires squealing, doorknobs turning or people walking. If the sound editor wants to communicate that a driver is in a hurry to leave, he will cut the sound of tires squealing when the car accelerates from a stop; even if the car is on a dirt road, the effect will work if the audience is dramatically engaged. If a character is afraid of someone on the other side of a door, the turning of the doorknob can take a second or more, and the mechanism of the knob can possess dozens of clicking parts. A skillful Foley artist can make someone walking calmly across the screen seem terrified simply by giving the actor a different gait.

Techniques

In music and film/television production, typical effects used in recording and amplified performances are:

- *echo* - to simulate the effect of reverberation in a large hall or cavern, one or several delayed signals are added to the original signal. To be perceived as echo, the delay has to be of order 35 milliseconds or above. Short of actually playing a sound in the desired environment, the effect of echo can be implemented using either analog or digital methods. Analog echo effects are implemented using tape delays and/or spring reverbs. When large numbers of delayed signals are mixed over several seconds, the resulting sound has the effect of being presented in a large room, and it is more commonly called reverberation or reverb for short.

- *flanger* - to create an unusual sound, a delayed signal is added to the original signal with a continuously variable delay (usually smaller than 10 ms). This effect is now done electronically using DSP, but originally the effect was created by playing the same recording on two synchronized tape players, and then mixing the signals together. As long as the machines were synchronized, the mix would sound more-or-less normal, but if the operator placed his finger on the flange of one of the players (hence "flanger"), that machine would slow down and its signal would fall out-of-phase with its partner, producing a phasing effect. Once the operator took his finger off, the player would speed up until its tachometer was back in phase with the master, and as this happened, the phasing effect would appear to slide up the frequency spectrum. This phasing up-and-down the register can be performed rhythmically.

- *phaser* - another way of creating an unusual sound; the signal is split, a portion is filtered with an all-pass filter to produce a phase-shift, and then the unfiltered and filtered signals are mixed. The phaser effect was originally a simpler implementation of the flanger effect since delays were difficult to implement with analog equipment. Phasers are often used to give a "synthesized" or electronic effect to natural sounds, such as human speech. The voice of C-3PO from *Star Wars* was created by taking the actor's voice and treating it with a phaser.

- *chorus* - a delayed signal is added to the original signal with a constant delay. The delay has to be short in order not to be perceived as echo, but above 5 ms to be audible. If the delay is too short, it will destructively interfere with the un-delayed signal and create a flanging effect. Often, the delayed signals will be slightly pitch shifted to more realistically convey the effect of multiple voices.

- *equalization* - different frequency bands are attenuated or boosted to produce desired spectral characteristics. Moderate use of equalization (often abbreviated as "EQ") can be used to "fine-tune" the tone quality of a recording; extreme use of equalization, such as heavily cutting a certain frequency can create more unusual effects.

- *filtering* - Equalization is a form of filtering. In the general sense, frequency ranges can be emphasized or attenuated using low-pass, high-pass, band-pass or band-stop filters. Band-pass filtering of voice can simulate the effect of a telephone because telephones use band-pass filters.

- *overdrive* effects such as the use of a fuzz box can be used to produce distorted sounds, such as for imitating robotic voices or to simulate distorted radiotelephone traffic (e.g., the radio chatter between starfighter pilots in the science fiction film *Star Wars*). The most basic overdrive effect involves *clipping* the signal when its absolute value exceeds a certain threshold.

- *pitch shift* - similar to pitch correction, this effect shifts a signal up or down in pitch. For example, a signal may be shifted an octave up or down. This is usually applied to the entire signal, and not to each note separately. One application of pitch shifting is pitch correction. Here a musical signal is tuned to the correct pitch using digital signal processing techniques. This effect is ubiquitous in karaoke machines and is often used to assist pop singers who sing out of tune. It is also used intentionally for aesthetic effect in such pop songs as Cher's "Believe" and Madonna's "Die Another Day".

- *time stretching* - the opposite of pitch shift, that is, the process of changing the speed of an audio signal without affecting its pitch.

- *resonators* - emphasize harmonic frequency content on specified frequencies.

- *robotic voice effects* are used to make an actor's voice sound like a synthesized human voice.

- *synthesizer* - generate artificially almost any sound by either imitating natural sounds or creating completely new sounds.

- *modulation* - to change the frequency or amplitude of a carrier signal in relation to a predefined signal. Ring modulation, also known as amplitude modulation, is an effect made famous by Doctor Who's Daleks and commonly used throughout sci-fi.

- *compression* - the reduction of the dynamic range of a sound to avoid unintentional fluctuation in the dynamics. Level compression is not to be confused with audio data compression, where the amount of data is reduced without affecting the amplitude of the sound it represents.

- *3D audio effects* - place sounds outside the stereo basis

- *reverse echo* - a swelling effect created by reversing an audio signal and recording echo and/or delay whilst the signal runs in reverse. When played back forward the last echos are heard before the effected sound creating a rush like swell preceding and during playback.

3D Audio Effect

3D audio is the use of binaural sound systems to capture, process and play back audio waves. The goal of 3D audio is to provide the listener with an audio experience that mimics real life.

3D audio recordings are made with two microphones mounted inside a human-like head and place where the human's ears would be. The microphones capture sound simultaneously through two channels and software adjusts the recording to mimic the slight variations in signals that occur when each ear sends signals to the brain, a concept known as biomimicry. Unlike surround-sound, which requires the use of multiple external speakers to provide directional audio sources, 3D audio can only be experienced through headphones.

Audio holograms that can replicate a 360-degree field audio experience and adjust for the disparate arrival time and amplitude of signals caused ear placement, have practical applications for training simulations and educational applications. 3D audio technology is expected to play an important part in both virtual reality (VR) and augmented reality (AR). Binaural 3D audio is particularly well suited to these emerging technologies because recordings are somewhat easy to produce and unlike specialized VR headsets, which can be quite heavy, 3D audio can be consumed through ordinary stereo headsets.

As is often the case with emerging technologies, there is an abundance of proprietary technologies and a lack of standardization with regards to audio holography. Licensing for 3D audio recording and processing technoloies can be quite expensive.Vendors associated with 3D audio include Dolby, 3Dio, Auro, Dysonics, Ossic, RealSpace, Sennheiser and VisiSonics.

Complete 3D Positional Audio

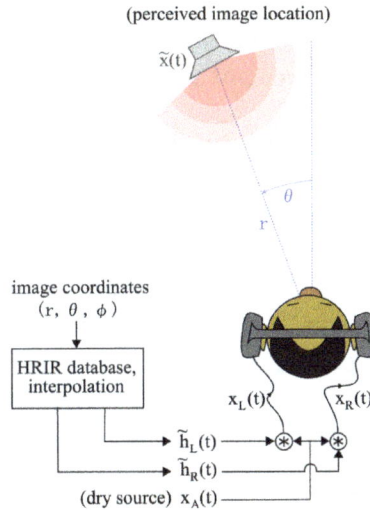

A sound is placed in the horizontal plane by processing the sound
with recorded head-related impulse responses.

Using head-related transfer functions and reverberation, the changes of sound on its way from the source (including reflections from walls and floors) to the listener's ear can be simulated. These effects include localization of sound sources behind, above and below the listener.

Some 3D technologies also convert binaural recordings to stereo recordings. Morrow-SoundTrue3D converts binaural, stereo, 5.1 and other formats to 8.1 single and multiple zone 3D sound experiences in realtime.

3D Positional Audio effects emerged in the 1990s in PC and Game Consoles.

3D audio techniques have also been incorporated in music and video-game style music video arts. The Audioscape research project, provides musicians with a real-time 3D audiovisual content authoring and rendering environment, suitable for live performance applications.

A site with animations and theory of a system using HRTF's to create 3D Audio: ISVR Virtual Acoustics.

True representation of the elevation level for 3D loudspeaker reproduction become possible by the Ambisonics and wave field synthesis (WFS) principle, MorrowSound True3D and A&G 3D-EST.

3-D Audio Presentations

Some amusement parks have created attractions based around the principles of 3-D audio. One example is *Sounds Dangerous!* at Disney's Hollywood Studios at the Walt Disney World Resort in Florida. Guests wear special earphones as they watch a short film starring comedian Drew Carey. At a point in the film, the screen goes dark while a 3-D audio sound-track immerses the guests in the ongoing story. To ensure that the effect is heard properly, the earphone covers are color-coded to indicate how they should be worn. This is not a generated effect but a binaural recording.

MorrowSoundTrue3D soundscapes include Torino Winter Olympics, ProFootball Hall of Fame, Great Lakes Children's Museum, NokiaWorld 2008 Barcelona, Denver Museum Nature and Science Gates Planetarium, New York Historical Society, Copenhagen International Theatre, Gallery Rachel Haferkamp Köln, Muu Gallery Helsinki, New Sounds New York, ZHDK Zurich, OKKO Design Stockholm, BAFTA Awards London, Collection of Diana Zlotnick Studio City, CA, as well as Ecsite, AAM, ASTC and IPS conventions. These range from single 8.1 to 64.3 True3D installations, some interactive.

Nick Cave's novel The Death of Bunny Munro was recorded in audiobook format using 3D audio.

The song "Propeller Seeds" by English artist Imogen Heap was recorded using 3d audio.

There has been developments in using 3D audio for DJ performances including the world's first Dolby Atmos event on 23rd Jan 2016 held at Ministry of Sound, London. The event was a showcase of a 3D audio DJ set performed by Hospital Records owner Tony Colman aka London Elektricity.

Other investigations included the Jago 3D Sound project which is looking at using Ambisonics combined with STEM music containers created and released by Native Instruments in 2015 for 3D nightclub sets.

Software

- Razer Surround
- Waves NX
- NuSpace Audio Muze and Zephyr
- Out Of Your Head

Flanging

Flanging is an audio process that combines two copies of the same signal, with the second delayed slightly, to produce a swirling effect. The process originated before digital effect boxes and computer editing were available. The effect, invented in the early 1950s by Les Paul and later used by artists such as Jimi Hendrix and The Beatles, was originally created using two tape recorders.

Here's how the nondigital process worked: While the original sound was being played from Tape Recorder #1, a second copy of the same audio material was played back from Tape Recorder #2. This process alone creates a hollow sound caused by the slight irregularities in the phase relationship of the audio waveforms. To get the flanging effect, the speed of the second recording was altered slightly. This was done most often by pressing a finger lightly on the tape reel's "flange", the large metal circle that surrounds and contains the tape on its hub. This created a time delay in addition to the phase differences, making the effect more pronounced.

Today, digital simulations of the process have replaced the flanging effect that was created using reel-to-reel tape recorders. The basic concept remains the same. The software or hardware device delays a copy of the source audio, but instead uses a low frequency oscillator (LFO) to vary the speed of the copy's playback. (The oscillator moves in the range of 1-20 cycles per seconds to get the effect.) Feeding the processed signal back into the device to be processed again can get a more intense effect.

Guitarist Les Paul invented flanging. He and Mary Ford first made it popular in the early 1950s. Les Paul also invented the solid body electric guitar and many sound techiques in use today. The first completely digital electronic flanging unit was the Delta Lab Research CompuEffectron, introduced in the 1970s.

Artificial Flanging

In the 1970s, advances in solid-state electronics made flanging possible using integrated circuit technology. Solid-state flanging devices fall into two categories: analog and digital. The flanging effect in most newer digital flangers relies on DSP technology. Flanging can also be accomplished using computer software.

The original tape-flanging effect sounds a little different from electronic and software recreations. Not only is the tape-flanging signal time-delayed, but response characteristics at different frequencies of the tape and tape heads introduced phase shifts into the signals as well. Thus, while the peaks and troughs of the comb filter are more or less in a linear harmonic series, there is a significant non-linear behaviour too, causing the timbre of tape-flanging to sound more like a combination of what came to be known as flanging and phasing.

"Barber Pole" Flanging

Also known as "infinite flanging", this sonic illusion is similar to the Shepard tone effect, and is equivalent to an auditory "barber pole". The sweep of the flanged sound seems to move in only one direction ("up" or "down") infinitely, instead of sweeping back-and-forth. While Shepard tones are created by generating a cascade of tones, fading in and out while sweeping the pitch either up or down, barber pole flanging uses a cascade of multiple delay lines, fading each one into the mix and fading it out as it sweeps to the delay time limit. The effect is available on various hardware and software effect systems.

Comparison with Phase Shifting

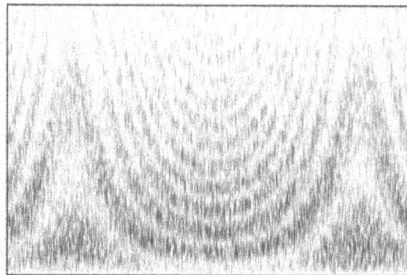

Flanging effect

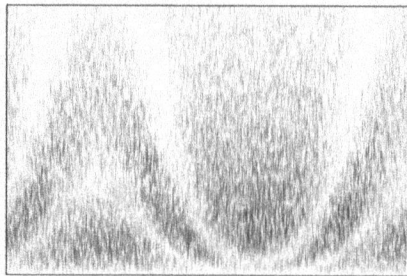

Phasing effect

Spectrograms of Phasing and Flanging effects

Flanging is one specific type of phase-shifting or "phasing". In phasing, the signal is passed through one or more all-pass filters which have non-linear phase response, and then added back to the original signal. This results in constructive and destructive interference that varies with frequency, giving a series of peaks and troughs in the frequency response of the system. In general, the position of these peaks and troughs do not occur in a harmonic series.

In contrast, flanging relies on adding the signal to a uniform time-delayed copy of itself, which results in an output signal with peaks and troughs which *are* in a harmonic series. Extending the comb analogy, flanging yields a comb filter with regularly spaced teeth, whereas phasing results in a comb filter with irregularly spaced teeth.

In both phasing and flanging, the characteristics (phase response and time delay, respectively) are generally varied in time, leading to an audible sweeping effect. To the ear, flanging and phasing sound similar, yet they are recognizable as distinct colorations. Commonly, flanging is referred to as having a "jet plane-like" characteristic. In order for the comb filter effect to be audible, the spectral content of the program material must be full enough within the frequency range of this moving comb filter to reveal the filter's effect. It is more apparent when it is applied to material with a rich harmonic content, and is most obvious when applied to a white noise or similar noise signal. If the frequency response of this effect is plotted on a graph, the trace resembles a comb, and so is called a comb filter.

Timeline

Flanging is a time-based effects unit that occurs when two identical signals are mixed together, but with one signal time-delayed by a small and gradually changing amount, usually smaller than 20 milliseconds. This produces a swept comb filter effect: peaks and notches are produced in the resultant frequency spectrum, related to each other in a linear harmonic series. Varying the time delay causes these to sweep up and down the frequency spectrum.

Part of the output signal is usually fed back to the input (a re-circulating delay line), producing a resonance effect which further enhances the intensity of the peaks and troughs. The phase of the fed-back signal is sometimes inverted, producing another variation on the flanging sound.

A *flanger* is a device dedicated to creating this sound effect.

Foley (Filmmaking)

Foley effects are sound effects added to the film during post production (after the shooting stops). They include sounds such as footsteps, clothes rustling, crockery clinking, paper folding, doors opening and slamming, punches hitting, glass breaking, etc. etc. In other words, many of the sounds that the sound recordists on set did their best to avoid recording during the shoot.

The boom operator's job is to clearly record the dialogue, and only the dialogue. At first glance it may seem odd that we add back to the soundtrack the very sounds the sound recordists tried to exclude. But the key word here is control. By excluding these sounds during filming and adding them in post, we have complete control over the timing, quality, and relative volume of the sound effects.

For example, an introductory shot of a biker wearing a leather jacket might be enhanced if we hear his jacket creak as he enters the shot - but do we really want to hear it every time he moves? By adding the foley sound fx in post, we can control its intensity, and fade it down once the dialogue begins. Even something as simple as boots on gravel can interfere with our comprehension of the dialogue if it is recorded too loudly. Far better for the actor to wear sneakers or socks (assuming their feet are off screen!) and for the boot-crunching to be added during Foley.

The technique is named after Jack Foley, who established the basic modern techniques still used today. Like most terms that are named in honour of a person, it is customary to spell Foley with a capital "F".

Uses

Foley complements or replaces sound recorded on set at the time of the filming, known as field recording. The soundscape of most films uses a combination of both. A Foley artist is the person who creates this sound art. Foley artists use creativity to make viewers believe that the sound effects are actually real. The viewers should not be able to realize that the sound was not actually part of the filming process itself. Foley sounds are added to the film in post production after the film has been shot. The need for replacing or enhancing sounds in a film production arises from the fact that, very often, the original sounds captured during shooting are obstructed by noise or are not convincing enough to underscore the visual effect or action. For example, fist-fighting scenes in an action movie are usually staged by the stunt actors and therefore do not have the actual sounds of blows landing. Crashes and explosions are often added or enhanced at the post-production stage. The desired effect is to add back to the original soundtrack the sounds that were intended to be excluded during recording. By excluding these sounds during field recording, and then adding them back into the soundtrack during post-production, the editors have complete control over how each noise sounds, its quality, and the relative volume. Foley effects add depth and realism to the audio quality for multimedia sources, and simplify the synchronizing of sounds that would otherwise be tedious or downright impossible to manage.

Foley artists review the film as it runs to figure out what sounds they need to achieve the desired sound and results. Once they gather the material and prepare for use, they practice the sounds. When they accomplish the desired sound, they watch the film and add in the sound effects at the same time. This is similar to the way actors re-record dialogue, lip-syncing to the video or film image.

Scenes where dialogue is replaced using dubbing also feature Foley sounds. Automatic dialogue replacement (ADR) is the process in which voice sounds are recorded in post production. This is done by a machine that runs the voice sounds with the film forward and backward to get the sound to run with the film. The objective of the ADR technique is to add sound effects into the film after filming, so the voice sounds are synchronized. Many sounds are not added at the time of filming, and microphones might not capture a sound the way the audience expects to hear it. The need for Foley rose dramatically when studios began to distribute films internationally, dubbed in foreign languages. As dialogue is replaced, all sound effects recorded at the time of the dialogue are lost as well.

Creation

Foley is created by the sound artist mimicking the actual sound source in a recording studio. Often there are many little sound effects that happen within any given scene of a movie. The process of recording them all can be tedious and time-consuming.

Foley art can be broken down into three main categories — feet, moves, and specifics.

Feet

The category entails the sound of footsteps. To make the sound of walking down a staircase, Foley artists stomp their feet on a marble slab while watching the footage. Foley studios carry many different types of shoes and several different types of floors to create footstep sounds. These floors, known as Foley Pits, vary from marble squares to gravel and rock pits. Creating just the right sound of footsteps can greatly enhance the feel of a scene. Foley Artists are often referred to as "Foley Walkers" or "Steppers" when working in the "feet" subset of Foley.

Moves

The "moves" category makes up many of the more subtle sounds heard in films, for example, the swishing of clothing when two actors walk past each other. This sound is created by rubbing two pieces of the same material together near the microphone at the same rate that the actor's legs cross. Cloth is not always used and tends to be recorded at the discretion of the dubbing mixer who ultimately controls the final outcome of the audio post-production process.

Specifics

Foley can also include other sounds, such as doors closing and doorbells ringing; however, these tend to be done more efficiently using stock sound effects, arranged by sound editors.

Foley effects help the viewer judge the size of a space. For example, a large hall has strong reverberation, while a small room may have only slight reverberation. Open outdoor spaces usually have no reverberation at all.

Common tricks

- Corn starch in a leather pouch makes the sound of snow crunching.
- A pair of gloves sounds like bird wings flapping.
- An arrow or thin stick makes a whoosh.
- An old chair makes a controllable creaking sound.
- A water-soaked rusty hinge when placed against different surfaces makes a

creaking sound. Different surfaces change the sound considerably.

- A heavy staple gun combined with other small metal sounds make good gun noises.

- A metal rake makes the rattle/squeak sound of chain-link fence; it can also make a metallic screech when dragged across a piece of metal.

- A heavy car door and fender can create most of the car sounds needed, but having a whole car in the studio is better.

- Burning plastic garbage bags cut into strips makes a realistic sounding candle or soft non-crackling fire when the bag melts and drips to the ground.

- ¼-inch audio tape balled up sounds like grass or brush when walked upon.

- Gelatin and hand soap make squishing noises.

- Frozen romaine lettuce makes bone or head injury noises.

- Coconut shells cut in half and stuffed with padding makes horse hoof noises; this is parodied in *Monty Python and the Holy Grail* and in *Shrek the Third*.

- Cellophane creates crackling fire effects.

- A selection of wooden and metal doors is needed to create all sorts of door noises, but also can be used for creaking boat sounds.

- Acorns, small apples and walnuts on a wooden parquet surface can be used for bones breaking

- Canned dog food can be used for alien pod embryo expulsions and monster vocalizations.

Soundstage & Acoustics

To capture a good recording, it is essential that your sound has sonic depth, width, and height. These spatial attributes are the difference between hi-fi and lo-fi recordings.

A typical room in an average sized home is going to be acoustically inferior to a larger room designed with acoustics in mind. Creating your own foley soundstage can aid you in squeezing that dimensional sound out of an average to smaller sized room. Additionally, the distortion of the depth, width, and height can be minimized by incorporating sound diffusion and sound absorption to nullify the room boundaries. The goal is to eliminate early reflections while maintaining a three-dimensional balance. That being said, constructing a soundstage is serious business.

Microphone Selection & Placement

Microphone selection is crucial to making foley audio "fit" with audio recorded on location. For interior scenes, a Neumann KM185, Oktava mk-012, Audix SCX1/HC or any hyper-cardiod condenser mic will capture good audio. Sensitive microphones are great at picking up subtle nuances in certain sound effects.

When matching foley for an outdoor scene, you may have better luck using a shotgun microphone similar to what is used on location shoots. There may even be some times where a large-diaphram condenser microphone is the way to go. By all means, use your ears to decide!

Proximity and placement of the microphone in relation to the source of sound greatly affects how the foley is recorded. Close-up shots may require closer microphone placement, or put distance between the mic and sound source if you desire more room sound. Always remember to experiment with different microphone positions.

Modern Foley Artists

Today, foley artists create entire atmospheres. They record the sounds of characters, props, and even clothing. Gary Hecker is a veteran foley artist who works on many of today's blockbuster films. He started his professional career in the 1970s and 80s, working on films like The Exorcist, Friday the 13th, The Empire Strikes Back, and Back to the Future.

References

- Atti, Andreas Spanias, Ted Painter, Venkatraman (2006). Audio signal processing and coding ([Online-Ausg.] ed.). Hoboken, NJ: John Wiley & Sons. p. 464. ISBN 0-471-79147-4

- Emmanuel Deruty; Damien Tardieu (January 2014). "About Dynamic Processing in Mainstream Music". Journal of the Audio Engineering Society. Retrieved 2014-06-06

- Jonathan Driedger and Meinard Müller (2016). "A Review of Time-Scale Modification of Music Signals". Applied Sciences. 6 (2): 57. doi:10.3390/app6020057

- Follansbee, Joe (2006). Hands-on Guide to Streaming Media: An Introduction to Delivering On-Demand Media. Focal Press. p. 84. ISBN 9780240808635

- Jens Hjortkjær; Mads Walther-Hansen (January 2014). "Perceptual Effects of Dynamic Range Compression in Popular Music Recordings". Journal of the Audio Engineering Society

- Esben Skovenborg (April 2012). "Loudness Range (LRA) – Design and Evaluation". AES 132nd Convention. Retrieved 2014-10-25

- Reese, David; Gross, Lynne; Gross, Brian (2009). Audio Production Worktext: Concepts, Techniques, and Equipment. Focal Press. p. 149. ISBN 0-240-81098-8

- "2010-2011 Tony Award Rules" (PDF). American Theatre Wing. Archived from the original (PDF) on 19 March 2012. Retrieved 26 December 2011

- Dugan, Dan (December 1969). "A New Music and Sound Effects System for Theatrical Productions". Journal of the Audio Engineering Society. Audio Engineering Society. 17 (6): 666–670. Retrieved March 23, 2011

- Serrà, J; Corral, A; Boguñá, M; Haro, M; Arcos, JL (26 July 2012). "Measuring the Evolution of Contemporary Western Popular Music". Scientific Reports. 2: 521. arXiv:1205.5651 . Bibcode:2012NatSR...2E.521S. doi:10.1038/srep00521. PMC 3405292 . PMID 22837813

3

Video Signals

Video signals can be both analog and digital. This chapter discusses in detail the different types of video signals, such as component video and composite video, as well as the aspects of video quality, video scaler, video file, coding format, etc.

Video started life as an analogue signal and remained so for over 50 years. However, digital television via satellite, cable and terrestrial transmission is now rapidly replacing analogue. Digital signals are less expensive to transmit because more information can be carried within a given bandwidth using digital multiplex and compression techniques. Digital signals also hold a significant advantage when it comes to video recording, editing and reproduction: whereas an analogue recording degrades with each generation of copying, the digital equivalent can nearly maintain the original quality after numerous copies. On initial analysis, picture quality is also superior to analogue with no apparent interference, ghosting or other problems. On closer inspection, however, errors can be detected, especially on rapidly moving pictures e.g. a football match.

Analogue Video

Analogue video is commonly distributed as a composite signal, an almost universal connection between video cameras, VCR/DVD players and video monitors. When superimposed on a radio frequency carrier it forms the aerial signal transmitted to homes. It is also the most usual input connection found on a videoconference picture monitor. This and other signal types are described below:

- The composite signal: The composite signal is composed of three parts: the black and white information (Luminance), the colour information (Chrominance) and the synchronisation (Synch) signals which ensure that the displayed pictures stay in close time synchrony with the transmission source. A problem with composite signals is that the three elements have to be coded to enable them to combine, but in the picture monitor these have to be decoded in order to display an image. These coding/decoding processes introduce unwanted noise and distortion. Three different coding systems are used worldwide. These systems – PAL, NTSC and SECAM – are all mutually incompatible. The NTSC (the National Television Standards Committee) system is used in North and South America and Japan. SECAM (SEquentiel Couleur Avec Memoire) is used in France and Eastern Europe, and PAL (Phase Alternating Line) is used in the UK and the rest of Europe.

- S-Video/YC: To reduce coding/decoding distortion the TV signal can be transmitted as an S-Video or YC signal. S-Video has two separate parts, Luminance (Y) and Chrominance (C), and so requires two separate connection channels between equipment in the chain. This involves less signal processing (decoding) in the picture monitor, which means less noise/distortion and thus a better picture.

- Alternatives: To reduce decoding noise even further the TV information may be transmitted as three signals, with a luminance channel (Y) and two colour component or colour difference channels, Red-Y (R-Y) and Blue-Y (B-Y). This minimises processing in the display monitor but requires three connection paths. An alternative method transmits Red (R), Green (G) and Blue (B), as separate components.

- SCART: The popular SCART interface includes three separate R G B channels together with composite and stereo sound signals, with the connected devices choosing the most appropriate video connection.

Digital Video Formats

Analogue video signals degrade when material is recorded or distributed. To overcome this, digital signals are now used throughout broadcasting. Another important advantage of digital signals is that massive compression is possible.

- CCIR-601/4:2:2: CCIR-601 was one of the first high quality digital standards to be introduced. It is also known as 4:2:2 or Y Cr Cb. It comprises Luminance (Y) and two Chrominance components (Cr and Cb) but as it requires a very wide bandwidth for transmission (around 166Mbit/s), it is rarely found outside broadcast studio environments.

- 4:2:0: To reduce the required bandwidth, and thus cost, other formats were developed, including 4:2:0. This has the same picture rate and luminance resolution as 4:2:2 but a reduced colour resolution, which is imperceptible to human eyes, as the human eye is much more sensitive to the luminance signal than to the chrominance signal. It is important as it forms the basis for the MPEG-2 (a video compression standard) form of coding used extensively for distributing digital television including SKY and Freeview.

- SIF and CIF: For less demanding applications SIF (Source Intermediate Format) was introduced. This has reduced frame rate and chrominance resolutions. An even lower quality format – CIF (Common Intermediate Format) which is a cross between the US and UK SIF formats – is used in videoconferencing. Other formats found in videoconferencing include the Quarter (QCIF), 4xCIF and 16CIF (for still images). QCIF has the lowest resolution and frame rate and is the base line format used within IP and ISDN conferencing for compatibility. A big advantage of CIF and its derivatives is that they are independent of origination television standards (PAL, NTSC etc.). This allows communication from the UK to the USA without standards conversion.

Characteristics of Video Streams

Number of Frames per Second

Frame rate, the number of still pictures per unit of time of video, ranges from six or eight frames per second (*frame/s*) for old mechanical cameras to 120 or more frames per second for new professional cameras. PAL standards (Europe, Asia, Australia, etc.) and SECAM (France, Russia, parts of Africa etc.) specify 25 frame/s, while NTSC standards (USA, Canada, Japan, etc.) specify 29.97 frame/s. Film is shot at the slower frame rate of 24 frames per second, which slightly complicates the process of transferring a cinematic motion picture to video. The minimum frame rate to achieve a comfortable illusion of a moving image is about sixteen frames per second.

Interlaced vs Progressive

Video can be interlaced or progressive. In progressive scan systems, each refresh period updates all scan lines in each frame in sequence. When displaying a natively progressive broadcast or recorded signal, the result is optimum spatial resolution of both the stationary and moving parts of the image. Interlacing was invented as a way to reduce flicker in early mechanical and CRT video displays without increasing the number of complete frames per second. Interlacing retains detail while requiring lower bandwidth compared to progressive scanning.

In interlaced video, the horizontal scan lines of each complete frame are treated as if numbered consecutively, and captured as two *fields*: an *odd field* (upper field) consisting of the odd-numbered lines and an *even field* (lower field) consisting of the even-numbered lines. Analog display devices reproduce each frame, effectively doubling the frame rate as far as perceptible overall flicker is concerned. When the image capture device acquires the fields one at a time, rather than dividing up a complete frame after it is captured, the frame rate for motion is effectively doubled as well, resulting in smoother, more lifelike reproduction of rapidly moving parts of the image when viewed on an interlaced CRT display.

NTSC, PAL and SECAM are interlaced formats. Abbreviated video resolution specifications often include an *i* to indicate interlacing. For example, PAL video format is often described as *576i50*, where *576* indicates the total number of horizontal scan lines, *i* indicates interlacing, and *50* indicates 50 fields (half-frames) per second.

When displaying a natively interlaced signal on a progressive scan device, overall spatial resolution is degraded by simple line doubling—artifacts such as flickering or "comb" effects in moving parts of the image which appear unless special signal processing eliminates them. A procedure known as deinterlacing can optimize the display of an interlaced video signal from an analog, DVD or satellite source on a progressive scan device such as an LCD television, digital video projector or plasma panel. Deinterlacing cannot, however, produce video quality that is equivalent to true progressive scan source material.

Aspect ratio

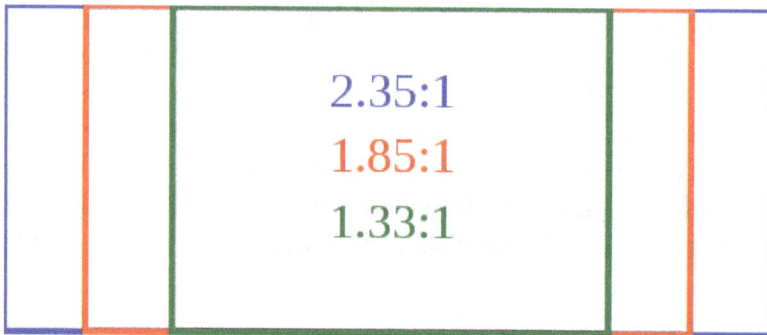

2.35:1

1.85:1

1.33:1

Comparison of common cinematographyand traditional television (green) aspect ratios

Aspect ratio describes the proportional relationship between the width and height of video screens and video picture elements. All popular video formats are rectangular, and so can be described by a ratio between width and height. The ratio width to height for a traditional television screen is 4:3, or about 1.33:1. High definition televisions use an aspect ratio of 16:9, or about 1.78:1. The aspect ratio of a full 35 mm film frame with soundtrack (also known as the Academy ratio) is 1.375:1.

Pixels on computer monitors are usually square, but pixels used in digital video often have non-square aspect ratios, such as those used in the PAL and NTSC variants of the CCIR 601 digital video standard, and the corresponding anamorphic widescreen formats. The 720 by 480 pixel raster uses thin pixels on a 4:3 aspect ratio display and fat pixels on a 16:9 display.

The popularity of viewing video on mobile phones has led to the growth of vertical video. Mary Meeker, a partner at Silicon Valley venture capital firm Kleiner Perkins Caufield & Byers, highlighted the growth of vertical video viewing in her 2015 Internet Trends Report – growing from 5% of video viewing in 2010 to 29% in 2015. Vertical video ads like Snapchat's are watched in their entirety nine times more frequently than landscape video ads.

Color Model and Depth

The color model the video color representation and maps encoded color values to visible colors reproduced by the system. There are several such representations in common use: YIQ is used in NTSC television, YUV is used in PAL television, YDbDr is used by SECAM television and YCbCr is used for digital video.

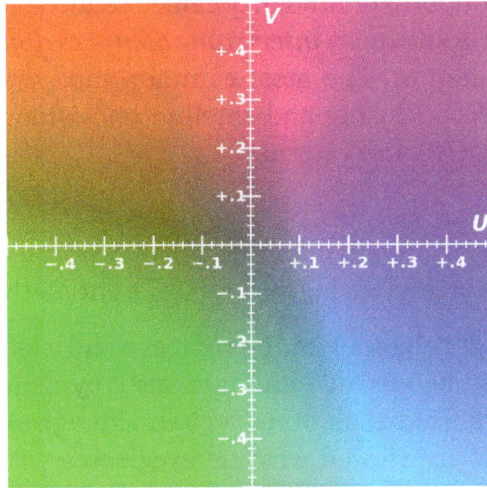

Example of U-V color plane, Y value=0.5

The number of distinct colors a pixel can represent depends on color depth expressed in the number of bits per pixel. A common way to reduce the amount of data required in digital video is by chroma subsampling (e.g., 4:4:4, 4:2:2, etc.). Because the human eye is less sensitive to details in color than brightness, the luminance data for all pixels is maintained, while the chrominance data is averaged for a number of pixels in a block and that same value is used for all of them. For example, this results in a 50% reduction in chrominance data using 2 pixel blocks (4:2:2) or 75% using 4 pixel blocks (4:2:0). This process does not reduce the number of possible color values that can be displayed, but it reduces the number of distinct points at which the color changes.

Video Quality

Video quality can be measured with formal metrics like Peak signal-to-noise ratio (PSNR) or through subjective video quality assessment using expert observation. Many subjective video quality methods are described in the ITU-T recommendation BT.500. One of the standardized method is the *Double Stimulus Impairment Scale* (DSIS). In DSIS, each expert views an *unimpaired* reference video followed by an *impaired* version of the same video. The expert then rates the *impaired* video using a scale ranging from "impairments are imperceptible" to "impairments are very annoying".

Video Compression Method (Digital Only)

Uncompressed video delivers maximum quality, but with a very high data rate. A variety of methods are used to compress video streams, with the most effective ones using a group of pictures(GOP) to reduce spatial and temporal redundancy. Broadly speaking, spatial redundancy is reduced by registering differences between parts of a single frame; this task is known as *intraframecompression* and is closely related to image compression. Likewise, temporal redundancy can be reduced by registering differences between frames; this task is known as *interframe compression*, including motion compensation and other techniques. The most common modern compression standards are MPEG-2, used for DVD, Blu-ray and satellite television, and MPEG-4, used for AVCHD, Mobile phones (3GP) and Internet.

Stereoscopic

Stereoscopic video can be created using several different methods:

- Two channels: a right channel for the right eye and a left channel for the left eye. Both channels may be viewed simultaneously by using light-polarizing filters 90 degrees off-axis from each other on two video projectors. These separately polarized channels are viewed wearing eyeglasses with matching polarization filters.

- One channel with two overlaid color-coded layers. This left and right layer technique is occasionally used for network broadcast, or recent "anaglyph" releases of 3D movies on DVD. Simple Red/Cyan plastic glasses provide the means to view the images discretely to form a stereoscopic view of the content.

- One channel with alternating left and right frames for the corresponding eye, using LCD shutter glasses that read the frame sync from the VGA Display Data Channel to alternately block the image to each eye, so the appropriate eye sees the correct frame. This method is most common in computer virtual reality applications such as in a Cave Automatic Virtual Environment, but reduces effective video framerate to one-half of normal (for example, from 120 Hz to 60 Hz).

Blu-ray Discs greatly improve the sharpness and detail of the two-color 3D effect in color-coded stereo programs.

Formats

Different layers of video transmission and storage each provide their own set of formats to choose from.

For transmission, there is a physical connector and signal protocol ("video connection standard" below). A given physical link can carry certain "display standards" that specify a particular refresh rate, display resolution, and color space.

Many analog and digital recording formats are in use, and digital video clips can also be stored on a computer file system as files, which have their own formats. In addition to the physical format used by the data storage device or transmission medium, the stream of ones and zeros that is sent must be in a particular digital *video compression format*, of which a number are available.

Analog Video

Analog video is a video signal transferred by an analog signal. An analog color video signal contains luminance, brightness (Y) and chrominance (C) of an analog television image. When combined into one channel, it is called composite video as is the case, among others with NTSC, PAL and SECAM.

Analog video may be carried in separate channels, as in two channel S-Video (YC) and multi-channel component video formats.

Analog video is used in both consumer and professional television production applications.

Composite video (single channel RCA)   S-Video (2-channel YC)   Component video (3-channel RGB)   TRRC   D-Terminal

Digital Video

Digital video signal formats with higher quality have been adopted, including serial digital interface (SDI), Digital Visual Interface (DVI), High-Definition Multimedia Interface (HDMI) and DisplayPortInterface, though analog video interfaces are still used and widely available. There exist different adaptors and variants.

Serial digital interface(SDI)   Digital Visual Interface(DVI)   HDMI   DisplayPort

Transport Medium

Video can be transmitted or transported in a variety of ways. Wireless broadcast as an analog or digital signal. Coaxial cable in a closed circuit system can be sent as analog interlaced 1 volt peak to peak with a maximum horizontal line resolution up to 480.

Broadcast or studio cameras use a single or dual coaxial cable system using a progressive scan format known as SDI serial digital interface and HD-SDI for High Definition video. The distances of transmission are somewhat limited depending on the manufacturer the format may be proprietary. SDI has a negligible lag and is uncompressed. There are initiatives to use the SDI standards in closed circuit surveillance systems, for Higher Definition images, over longer distances on coax or twisted pair cable. Due to the nature of the higher bandwidth needed, the distance the signal can be effectively sent is a half to a third of what the older interlaced analog systems supported.

Digital Television

New formats for digital television broadcasts use the MPEG-2 video coding format and include:

- ATSC – United States, Canada, Mexico, Korea
- Digital Video Broadcasting (DVB) – Europe
- ISDB – Japan
    - o ISDB-Tb – uses the MPEG-4 video coding format – Brazil, Argentina
- Digital Multimedia Broadcasting (DMB) – Korea

Analog Television

Analog television broadcast standards include:

- FCS – USA, Russia; obsolete
- MAC – Europe; obsolete
- MUSE – Japan
- NTSC – United States, Canada, Japan
- PAL – Europe, Asia, Oceania
    - o PAL-M – PAL variation. Brazil, Argentina
    - o PALplus – PAL extension, Europe
- RS-343 (military)
- SECAM – France, former Soviet Union, Central Africa

An analog video format consists of more information than the visible content of the frame. Preceding and following the image are lines and pixels containing synchronization information or a time delay. This surrounding margin is known as a blanking interval or blanking region; the horizontal and vertical front porch and back porch are the building blocks of the blanking interval.

Recording Formats before Video Tape

- Phonovision

- Kinescope

Analog Tape Formats

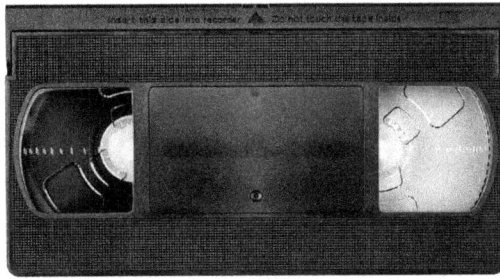

A VHS video cassette tape.

In approximate chronological order. All formats listed were sold to and used by broadcasters, video producers or consumers; or were important historically (VERA).

- 2" Quadruplex videotape (Ampex 1956)

- VERA (BBC experimental format ca. 1958)

- 1" Type A videotape (Ampex)

- 1/2" EIAJ (1969)

- U-matic 3/4" (Sony)

- 1/2" Cartrivision (Avco)

- VCR, VCR-LP, SVR

- 1" Type B videotape (Robert Bosch GmbH)

- 1" Type C videotape (Ampex, Marconi and Sony)

- Betamax (Sony)

- VHS (JVC)

- Video 2000 (Philips)

- 2" Helical Scan Videotape (IVC)

- 1/4" CVC (Funai)

- Betacam (Sony)

- HDVS (Sony)

- Betacam SP (Sony)

- Video8 (Sony) (1986)

- S-VHS (JVC) (1987)

- VHS-C (JVC)

- Pixelvision (Fisher-Price)

- UniHi 1/2" HD (Sony)

- Hi8 (Sony) (mid-1990s)

- W-VHS (JVC) (1994)

Digital Tape Formats

- Betacam IMX (Sony)

- D-VHS (JVC)

- D-Theater

- D1 (Sony)

- D2 (Sony)

- D3

- D5 HD

- D6 (Philips)

- Digital-S D9 (JVC)

- Digital Betacam (Sony)

- Digital8 (Sony)

- DV (including DVC-Pro)

- HDCAM (Sony)

- HDV

- ProHD (JVC)

- MicroMV

- MiniDV

Optical Disc Storage Formats

- Blu-ray Disc (Sony)
- China Blue High-definition Disc (CBHD)
- DVD (was Super Density Disc, DVD Forum)
- Professional Disc
- Universal Media Disc (UMD) (Sony)

Discontinued

- Enhanced Versatile Disc (EVD, Chinese government-sponsored)
- HD DVD (NEC and Toshiba)
- HD-VMD
- Capacitance Electronic Disc
- Laserdisc (MCA and Philips)
- Television Electronic Disc (Teldec)) and (Telefunken)
- VHD (JVC)

Digital Encoding Formats

- CCIR 601 (ITU-T)
- H.261 (ITU-T)
- H.263 (ITU-T)
- H.264/MPEG-4 AVC (ITU-T + ISO)
- H.265
- M-JPEG (ISO)
- MPEG-1 (ISO)
- MPEG-2 (ITU-T + ISO)
- MPEG-4 (ISO)
- Ogg-Theora
- VP8-WebM
- VC-1 (SMPTE)

Clean Feed (Television)

A video feed without any graphics overlaid and usually with only natural sound audio.

Distribution master tapes from programme-makers are usually delivered with the entire programme, followed by a section of 'textless material': anything which would have a graphic on it, without the graphics (and usually in silence). The main programme material will have the final mix audio on channels 1 and 2, and a 'music and effects' mix (ie no voiceover) on 3 and 4.

A clean feed from a broadcaster will normally be just the programmes, with no trails, interstitials, or DVEs. In the BBC's case, long blank gaps between programmes are filled with a slide of the BBC logo on some clouds.

A clean feed is a signal which has not come from the main output of the video switcher, such as the output of a vision mixer before the downstream keyer stage - the clean feed is identical to the main program output but without any captions keyed into it. Modern production equipment can actually put different keys on multiple outputs, allowing them to go to the clean feed or not. The most sophisticated vision mixers (or production switchers, according to the American nomenclature) can generate a clean feed output for any of their mix/effects (ME) buses.

The term "clean feed" is also used to refer to backhaul feeds of television programming sent via communication satellite or other transport (such as a national fiber-optic network) sent from another TV station or remote television production truck on-location, which does not carry any television advertisements or break bumpers, or in some cases, lower-third graphics or superimposed chyron text.

Component Video

Component video is a video signal that is transmitted in several parts in order to optimize quality. It is common in analog video transmission and is often paired with an accompanying audio signal. The different parts of component video signals generally consist of chroma (color) or luma (light) information, although sometimes this varies. Component video contrasts with composite video, which combines all segments of a video signal into a single channel.

Analog video developed concurrently with the advent of the television, and component video came with the rise of color TV in the 1950s. It provided a medium through which motion picture information could transmit reliably and efficiently, and still exists today in media display environments and setups wherein analog transmission is employed.

Component video cables generally have multiple connectors or pins on either side of their throughput, since they transmit signals separately from one another in order to achieve the best possible quality on the display end. One of the most common component video cable formats is RCA, which usually features three different connectors (for green, blue and red color information), with many of them also being paired with one or two more connectors for audio.

Analog Component Video

Reproducing a video signal on a display device (for example, a cathode ray tube (CRT)) is a straightforward process complicated by the multitude of signal sources. DVD, VHS, computers and video game consoles all store, process and transmit video signals using different methods, and often each will provide more than one signal option. One way of maintaining signal clarity is by separating the components of a video signal so that they do not interfere with each other. A signal separated in this way is called "component video". S-Video, RGB and YP_BP_R signals comprise two or more separate signals, and thus are all component-video signals. For most consumer-level video applications, the common 3 cable system using BNC or RCA connectors analog component video was used. Typical resolutions (in lines) are 480i (DVD) and 576i (US and Japan broadcast analog TV). For personal computer displays the 15 pin DIN connector (IBM VGA) provided screen resolutions including 640x480, 800x600, 1024x768, 1152x864, 1280x1024 and much larger.

RGB Analog Component Video

A 15-pin VGA connector for a personal computer   A 21-pin SCART connector

The various RGB (red, green, blue) analog component video standards (e.g., RGBS, RGBHV, RGsB) use no compression and impose no real limit on color depth or resolution, but require large bandwidth to carry the signal and contain a lot of redundant data since each channel typically includes much of the same black-and-white image. Most modern computers offer this signal via a VGA port. Many televisions, especially in Europe, utilize RGB via the SCART connector. All arcade games, other than early vector and black-and-white games, use RGB monitors.

In addition to the red, green and blue color signals, RGB requires two additional signals to synchronize the video display. Several methods are used:

- composite sync, where the horizontal and vertical signals are mixed together on a separate wire (the S in RGBS)

- separate sync, where the horizontal and vertical are each on their own wire (the H and V in RGBHV; also the acronym HD/VD, meaning *horizontal deflection/ vertical deflection*, is used)

- sync on green, where a composite sync signal is overlaid on the wire used to transport the green signal (SoG, Sync on G, or RGsB).

- sync on red or sync on blue, where a composite sync signal is overlaid on either the red or blue wire

- sync on composite where the signal normally used for composite video is used alongside the RGB signal only for the purposes of sync.

- sync on luma, where the Y signal from S-Video is used alongside the RGB signal only for the purposes of sync.

Composite sync is common in the European SCART connection scheme (using pins 17 [ground] and 19 [composite-out] or 20 [composite-in]). RGBS requires four wires – red, green, blue and sync. If separate cables are used, the sync cable is usually colored yellow (as is the standard for composite video) or white.

Separate sync is most common with VGA, used worldwide for analog computer monitors. This is sometimes known as RGBHV, as the horizontal and vertical synchronization pulses are sent in separate channels. This mode requires five conductors. If separate cables are used, the sync lines are usually yellow (H) and white (V), yellow (H) and black (V), or gray (H) and black (V).

Sync on Green (SoG) is less common, and while some VGA monitors support it, most do not. Sony is a big proponent of SoG, and most of their monitors (and their PlayStation line of video game consoles) use it. Like devices that use composite video or S-video, SoG devices require additional circuitry to remove the sync signal from the green line. A monitor that is not equipped to handle SoG will display an image with an extreme green tint, if any image at all, when given a SoG input.

Sync on red and sync on blue are even rarer than sync on green, and are typically used only in certain specialized equipment.

Sync on composite, not to be confused with composite sync, is commonly used on devices that output both composite video and RGB over SCART. The RGB signal is used for color information, while the composite video signal is only used to extract the sync information. This is generally an inferior sync method, as this often causes checkerboards to appear on an image, but the image quality is still much sharper than standalone composite video.

Sync on luma is much similar to sync on composite, but uses the Y signal from S-Video instead of a composite video signal. This is sometimes used on SCART, since both composite video and S-Video luma ride along the same pins. This generally does not suffer from the same checkerboard issue as sync on composite, and is generally acceptable on devices that do not feature composite sync, such as the Sony PlayStation and some modded Nintendo 64 models.

Luma-based Analog Component Video

YP_BP_R component video out on a consumer electronics device

Further types of component analog video signals do not use separate red, green and blue components but rather a colorless component, termed luma, which provides brightness information (as in black-and-white video). This combines with one or more color-carrying components, termed chroma, that give only color information. Both the S-Video component video output (two separate signals) and the YP_BP_R component video output (three separate signals) seen on DVD players are examples of this method.

Converting video into luma and chroma allows for chroma subsampling, a method used by JPEG and MPEG compression schemes to reduce the storage requirements for images and video (respectively).

Many consumer TVs, DVD players, monitors, video projectors and other video devices use YP_BP_R output or input.

When used for connecting a video source to a video display where both support 4:3 and 16:9 display formats, the PAL television standard provides for signaling pulses that will automatically switch the display from one format to the other.

Connectors Used

- D-Terminal: Used mostly on Japanese electronics.

- Three BNC (professional) or RCA connectors (consumer): Typically colored green (Y), blue (P_B) and red (P_R).

- SCART used in Europe.

- Video In Video Out (VIVO): 9-pin Mini-DIN-connectors called "TV Out" in computer video cards, which usually include an adaptor for component RCA, composite RCA and 4-pin S-Video-Mini-DIN.

Digital Component Video

Digital component video makes use of single cables with signal lines/connector pins dedicated to digital signals, transmitting digital color space values allowing higher resolutions such as 480p, 576i, 576p, 720p, 1080i, and 1080p.

RGB component video has largely been replaced by modern digital formats, such as DisplayPort or Digital Visual Interface (DVI) digital connections, while home theater systems increasingly favor High-Definition Multimedia Interface (HDMI), which support higher resolutions, higher dynamic range, and can be made to support digital rights management. The demise of analog is largely due to screens moving to large flat digital panels as well as the desire for having a single cable for both audio and video but also due to a slight loss of clarity when converting from a digital media source to analogue and back again for a flat digital display, particularly when used at higher resolutions where analog signals are highly susceptible to noise.

International Standards

Examples of international component video standards are:

- RS-170 RGB (525 lines, based on NTSC timings, now EIA/TIA-343)

- RS-343 RGB (525, 625 or 875 lines)

- STANAG 3350 Analogue Video Standard (NATO military version of RS-343 RGB, now EIA-343A)

- CEA-770.3 High Definition TV Analog Component Video Interface Consumer Electronics Association

Component Versus Composite

In a composite signal, the luminance, Brightness (Y) signal and the chrominance, Color (C) signals are encoded together into one signal. When the color components are kept as separate signals, the video is called component analog video (CAV), which requires three separate signals: the luminance signal (Y) and the color difference signals (R-Y and B-Y).

Since component video does not undergo the encoding process, the color quality is noticeably better than composite video.

Component video connectors are not unique in that the same connectors are used for several different standards; hence, making a component video connection often does not lead to a satisfactory video signal being transferred. Many DVD players and TVs may need to be set to indicate the type of input/output being used, and if set incorrectly the image may not be properly displayed. Progressive scan, for example, is often not enabled by default, even when component video output is selected.

Composite Video

Composite video is a method in which the Color, B/W, and Luminance portions of an analog video signal are transferred together from a source to a video recording device (VCR, DVD recorder) or video display (TV, monitor, video projector). Composite video signals are analog and typically consists of 480i (NTSC)/576i (PAL) standard definition resolution video signals. Composite video, as applied in the consumer environment, is not designed to be used for transferring high definition analog or digital video signals.

The composite video signal format is also referred to as CVBS (Color, Video, Blanking, and Sync or Color, Video, Baseband, Signal), or YUV (Y = Luminance, U, and V = Color).

It must be pointed out that composite video is not the same as RF signal is transferred from an antenna or cable box to a TV's RF inputs using a Coaxial Cable - the signals are not the same. RF refers to Radio Frequency, which are signals transmitted over the air, or relayed through a cable or satellite box to the antenna input connection on a TV via a screw-on or push-on coaxial cable.

Signal Components

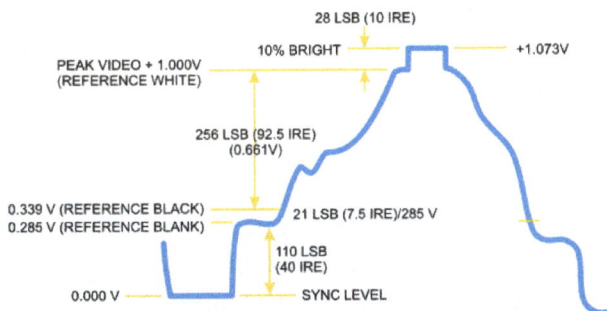

A composite video signal combines on one wire the video information required to recreate a color picture, as well as line and frame synchronization pulses. The color video signal is a linear combination of the *luminance* of the picture, and a modulated subcarrier carries the *chrominance* or color information, a combination of hue and saturation. Details of the encoding process vary between the NTSC, PAL and SECAM systems.

The frequency spectrum of the modulated color signal overlaps that of the baseband signal, and separation relies on the fact that frequency components of the baseband signal tend to be near harmonics of the horizontal scanning rate, while the color carrier is selected to be an odd multiple of half the horizontal scanning rate; this produces a modulated color signal that consists mainly of harmonic frequencies that fall between the harmonics in the baseband luma signal, rather than both being in separate continuous frequency bands alongside each other in the frequency domain. In other words, the combination of luma and chroma is indeed a frequency-division technique, but it is much more complex than typical frequency-division multiplexing systems like the one used to multiplex analog radio stations on both the AM and FM bands.

A gated and filtered signal derived from the color subcarrier, called the burst or color-burst, is added to the horizontal blanking interval of each line, excluding the vertical sync interval, as a synchronizing signal and amplitude reference for the chrominance signals. The burst signal is inverted in phase (180° out of phase) from the reference subcarrier.

Signal Modulation

Composite video can easily be directed to any broadcast channel simply by modulating the proper RF carrier wave with it. Most home analog video equipment record a signal in (roughly) composite format: LaserDiscs store a true composite signal, while consumer videotape formats (including VHS and Betamax) and lesser commercial and industrial tape formats (including U-Matic) use modified composite signals (generally known as *color-under*). On playback, these devices often give the user the option to output the baseband signal or to modulate it onto a VHF or UHF frequency compatible with a TV tuner (i.e., appearing on a selected TV channel). The professional television production uncompressed digital video videocassette format known as D-2 (video) directly records and reproduces standard NTSC composite video signals, using PCM encoding of the analog signal on the magnetic tape.

Standard Connectors

In home applications, the composite video signal is typically connected using an RCA connector (phono plug), normally yellow. It is often accompanied with red and white (or black) connectors for right and left audio channels respectively. BNC connectors and higher quality coaxial cable are often used in professional television studios and post-production applications. BNC connectors were also used for composite video connections on early home VCRs, often accompanied by either phono connectors or a 5-pin DIN connector for audio. The BNC connector, in turn post dated the PL-259 connector which featured on first generation VCRs.

In Europe, SCART connections are often used instead of RCA jacks (and to a lesser extent, S-Video), so where available, RGB is used instead of composite video with computers, video game consoles, and DVD players.

Video cables are 75 ohm impedance, low in capacitance. Typical values run from 52 pF/m for an HDPE-foamed dielectric precision video cable to 69 pF/m for a solid PE dielectric cable.

Modulators

Some devices that connect to a TV, such as VCRs, older video game consoles and home computers of the 1980s, output a composite signal. This may then be converted to RF with an external box known as an RF modulator that generates the proper carrier (often for channel 3 or 4 in North America, channel 36 in Europe). Sometimes this modulator was built into the product (such as video game consoles, VCRs, or the Atari, Commodore 64, or TRS-80 CoCo home-computers) and sometimes it was an external unit powered by the computer (in the case of the TI-99/4A or some Apple modulators) or with an independent power supply. In the United States, using an external RF modulator frees the manufacturer from obtaining FCC approval for each variation of a device. Through the early 1980s, electronics that output a television channel signal were required to meet the same shielding requirements as broadcast television equipment, thus forcing manufacturers such as Apple to omit an RF modulator, and Texas Instruments to have their RF modulator as an external unit, which they had certified by the FCC without mentioning they were planning to sell it with a computer. In Europe, while most countries used the same broadcast standard, there were different modulation standards (PAL-G versus PAL-I, for example), and using an external modulator allowed manufacturers to make a single product and easily sell it to different countries by changing the modulator. Video game consoles on the other hand were less of an issue with FCC approval because the circuitry was inexpensive enough to allow for channel 3/4 outputs.

Modern day devices with analog outputs have typically omitted channel 3 and 4 outputs in favor of composite and S-video outputs (or have switched to using HDMI or other digital formats) as composite and S-video have become more common as inputs for TVs. In addition, many TV sets sold these days no longer have analog television tuners and cannot accept channel 3/4. But because composite video has a well-established market for both devices that convert it to channel 3/4 outputs, as well as devices that convert things like VGA to composite, it has offered opportunities to repurpose older composite monitors for newer devices.

Demodulation Loss

The process of modulating RF with the original video signal, and then demodulating the original signal again in the TV, introduces several losses. This conversion also typically adds noise or interference to the signal as well. For these reasons, it is typically best to use composite connections instead of RF connections if possible. Almost all modern video equipment has at least composite connectors, so this typically is not a problem; however, older video equipment and some very low-end modern televisions have only RF input (essentially the antenna jack); while RF modulators are no longer common, they are still available to translate baseband signals for older equipment.

However, just as the modulation and demodulation of RF loses quality, the mixing of the various signals into the original composite signal does the same, causing a checkerboard video artifact known as dot crawl. Dot crawl is a defect that results from crosstalk due to the intermodulation of the chrominance and luminance components of the signal. This is usually seen when chrominance is transmitted with a high bandwidth, and its spectrum reaches into the band of the luminance frequencies. This has led to a proliferation of systems such as S-Video and component video to maintain the signals separately. Comb filters are also commonly used to separate signals, and eliminate artifacts, from composite sources.

Rear of Polish computer Elwro 800 Junior. DIN output called "MONITOR" is just a Composite Video (no color info, only black and white), Sound (mono) and Ground output in that form. Despite computer being designed (and mainly made of Eastern components) on East side of the Iron curtain it can be connected to modern equipment like VGA-Composite video adapters or monitors with Composite Video input, for example using a popular and simple chinch-DIN cable.

Aspect ratio in Composite Signal

When used for connecting a video source to a video display that supports both 4:3 and 16:9 display formats, the PAL and NTSC television standards provide for signaling pulses that will automatically switch the display from one format to the other. This is called widescreen signalling (WSS).

Extensions to the Composite Video Standard

Since TV screens hide the vertical blanking interval of a composite video signal and even crop the edges of the picture, extensions have been implemented by taking advantage of these unseen parts of the signal. Examples of these extensions include teletext, closed captioning, digital information regarding the show title, transmitting a set of reference colors that allows TV sets to automatically correct the hue maladjustments common with the NTSC color encoding system, etc.

Other extensions to the standard include S-video; S-video is an extension to the standard because it uses parallel signal paths for luminance and for chrominance (color), of which both of them can be connected to a composite video input but with either monochrome (luma), or uniform-luma color (chroma) unless merging the signal paths with a filter was done.

Problems

A TV or projector needs to separate the various bits of picture information (luma, chroma and sync) in order to re-create the picture, and this is the problem with composite video. It is impossible to separate all the information back to the original quality once it has been mixed together in the one cable. This means the picture is not as crisp and the colours not as defined as they could be.

Colorburst

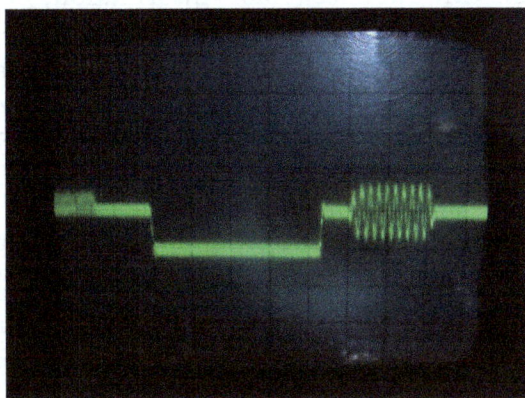

Horizontal sync and color burst of PAL videosignal

Colorburst is an analog video, composite video signal generated by a video-signal generator used to keep the chrominance subcarrier synchronized in a color television signal. By synchronizing an oscillator with the colorburst at the back porch (beginning) of each scan line, a television receiver is able to restore the suppressed carrier of the chrominance (color) signals, and in turn decode the color information. The most common use of colorburst is to genlock equipment together as a common reference with a vision mixer in a television studio using a multi-camera setup.

Explanation

In NTSC, its frequency is exactly $315/88 = 3.57954$[a] MHz with a phase of 180°. PAL uses a frequency of exactly 4.43361875 MHz, with its phase alternating between 135° and 225° from line to line. Since the colorburst signal has a known amplitude, it is sometimes used as a reference level when compensating for amplitude variations in the overall signal.

SECAM is unique in not having a colorburst signal, since the chrominance signals are encoded using FM rather than QAM, thus the signal phase is immaterial and no reference point is needed.

Rationale for NTSC Color Burst Frequency

The original black and white NTSC television standard specified a frame rate of 30 Hz and 525 lines per frame, or 15750 lines per second. The audio was frequency modulated 4.5 MHz above the video signal. Because this was black and white, the video consisted only of luminance (brightness) information. Although all of the space in between was occupied, the line-based nature of the video information meant that the luminance data was not spread uniformly across the frequency domain; it was concentrated at multiples of the line rate. Plotting the video signal on a spectrogram gave a signature that looked like the teeth of a comb or a gear, rather than smooth and uniform.

RCA discovered that if the chrominance (color) information, which had a similar spectrum, was modulated on a carrier that was a half-integer multiple of the line rate, its signal peaks would fit neatly between the peaks of the luminance data and interference was minimized. It was not eliminated, but what remained was not readily apparent to human eyes. (Modern televisions attempt to reduce this interference further using a comb filter.)

To provide sufficient bandwidth for the chrominance signal, yet interfere only with the highest-frequency (and thus least perceptible) portions of the luminance signal, a chrominance subcarrier near 3.6 MHz was desirable. $227.5 = 455/2$ times the line rate was close to the right number, and 455's small factors ($5 \times 7 \times 13$) make a divider easy to construct.

However, additional interference could come from the audio signal. To minimize interference there, it was similarly desirable to make the distance between the chrominance carrier frequency and the audio carrier frequency a half-integer multiple of the line rate. The sum of these two half-integers implies that the distance between the frequency of the luminance carrier and audio carrier must be an integer multiple of the line rate. However, the original NTSC standard, with a 4.5 MHz carrier spacing and a 15750 Hz line rate, did not meet this requirement: the audio was 285.714 times the line rate.

While existing black and white receivers could not decode a signal with a different audio carrier frequency, they could easily use the copious timing information in the video signal to decode a slightly slower line rate. Thus, the new color television standard reduced the line rate by a factor of 1.001 to 1/286 of the 4.5 MHz audio subcarrier frequency, or about 15734.2657 Hz. This reduced the frame rate to $30/1.001 \approx 29.9700$

Hz, and placed the color subcarrier at $227.5/286 = 455/572 = 35/44$ of the 4.5 MHz audio subcarrier.

Crystals

An NTSC or PAL television's color decoder contains a colorburst crystal oscillator.

Because so many analog color TVs were produced from the 1960s to the early 2000s, economies of scale drove down the cost of colorburst crystals, so they were often used in various other applications, such as oscillators for microprocessors or for amateur radio. (3.5795 MHz has since become a common QRP calling frequency in the 80-meter band, and it's doubled frequency of 7.159 MHz is a common calling frequency in the 40-meter band).

Color Framing

In video engineering, color framing refers to the color frame sequence of fields in a composite video signal through which the video frame timing and chrominance subcarrier signal timing—in particular, that of the color burst -- cycle through all possible phase relationships.

The exact nature of the color frame sequence depends on the video standard being used. In the case of the three main composite video standards, PAL video has an 8-field (4 frame) color frame sequence, and NTSC and SECAM both have 4-field (2 frame) color frame sequences.

Preserving the color framing sequence of video across edits and between channels in video effects was an important issue in early analog composite videotape editing systems, as cuts between different color sequences would cause jumps in subcarrier phase, and mixing two signals of different field dominance would result in color artifacts on the part of the signal that was not in sync with the output color frame sequence.

To help prevent these problems, SMPTE time code contains a color framing bit, which can be used to indicate that the video material the timecode refers to follows a standard convention regarding the synchronization of video time code and the color framing sequence. If the color framing bit was set in both types of material, the editing system could then always ensure that color framing was preserved by constraining edit decisions between input sources to keep the correct relationship between the timecode sequences, and hence the color framing sequences.

Color framing has become largely an issue of historical interest, first with the advent in the 1980s of digital composite video timebase correctors and frame stores, which could regenerate the color frame sequence of a composite signal at any phase, and later with analog component video editing and modern digital video systems, in which subcarrier phase is no longer relevant.

IRE (Unit)

A composite video signal is measured in IRE (Institute of Radio Engineers). It is a percentage of the total voltage. This is how it looks, if you have the courage to look, that is:

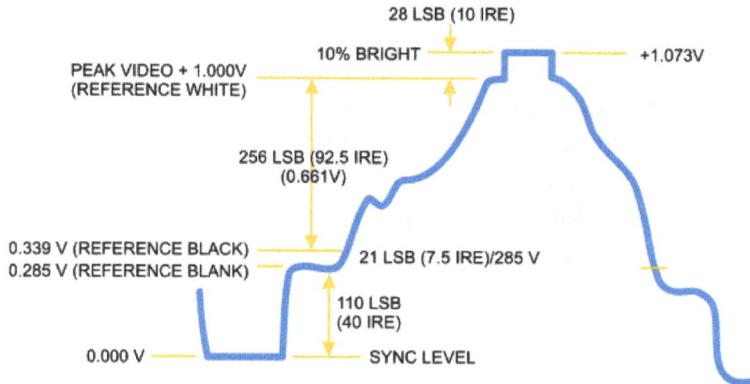

- 0 IRE is 0 volts is black.

- 100 IRE is white.

- There is also a sync pulse of 40 IRE that is sneaked in below 0.

- Therefore, the total IRE of a composite signal is 140 IRE, and the total voltage is about 1 V.

Signals are measured in percentages compared to 140 IRE. Since the total voltage is 1V, the actual image is contained within approximately 0.7 V or 700mV (milli Volts).

You can raise the amplitude of this signal, but the IRE must remain constant. PAL, NTSC and SECAM are all transferred this way.

For PAL, black is at 0 IRE and white is at 100 IRE.

For NTSC, black is at 7.25 IRE (US), and 0 IRE (Japan); and white is at 100 IRE.

Importance of IRE

IRE is like quality control, and it is one of those properties that directly measure the strength of a signal. Low IRE values could mean the signal isn't strong enough, or it doesn't contain enough information. In either case, poor IRE values are a cause for concern.

Video Quality

Video quality is the degree at which a given video suits the end-user's expectations. It is therefore a subjective notion. The proof is that a given video, viewed in the same conditions by different observers, can be judged very differently by these observers. One observer could say that the video quality is good while another one could say that the video quality is bad, and both would be right: because it depends on the video quality they expect and on the video quality they're used to have.

Another way to express video quality is to talk about visual annoyance because visual annoyance is often a bit easier to evaluate than video quality. Visual annoyance represents the level of difficulty you meet when trying to perform a given task (from "simple comfort" to face expressions detection or object recognition) in presence of video distortions (like encoding artifacts or decoding errors due to packets loss during transmission).

Video quality judgment and visual annoyance both come from visual perception and can be represented by the top-down approach below:

Video quality judgement

Visual annoyance

Visual perception

Video Quality should be Subjectively Assessed

Video quality is a subjective notion and therefore the best way to measure video quality is based on human observers. This way is called "subjective video quality assessment". It requires to use human observers who must score the video quality of videos they're watching during experiments called "video quality assessment tests".

However, in order to collect precise video quality measures, many recommendations have to be followed when realizing such subjective video quality assessment tests. Otherwise, the collected video quality scores often become useless (due to the lack of precision on the video quality scores).

These recommendations lead the subjective video quality assessment method to be very complicated. More, performing subjective video quality assessments tests is a long and expensive process.

Video Quality can also be Objectively Measured

Video quality is difficult to measure with the "subjective video quality assessment" method descrived above. Luckily, there is another way to measure video quality. This other way is called "objective video quality measurement".

It is based on the use of a computational method, an algorithm, called "metric" (or "video quality metric" or "video quality criterion") which produces values expressing video quality.

One of the fundamental properties required for a video quality metric is that it should produce objective video quality scores which are well correlated with subjective video quality scores collected from human observers during video quality assessment tests. From a practical point of view, a video quality metric is an algorithm able to score (on a scale) the video quality of a tested video which may have been distorted (usually due to encoding and/or transmission).

While computing video quality scores can be very easy, producing precise, meaningful and coherent video quality scores (which means video quality scores well correlated with subjective video quality scores given by human observers) is much more complicated and requires important research efforts on human visual perception, features extraction, distortions visibility and pooling of disortions into video quality scores.

The interest of these video quality metrics is that they enable to measure or monitor in real time the video quality of audio video services, in a fully automatic and repeatable manner. While video quality measurement is often used for equipment benchmarking (to buy the best equipment, the one which provides the best video quality), video quality monitoring solutions can trigger alerts when equipment fails to reach a given video quality level.

From Analog to Digital Video

Since the world's first video sequence was recorded and transmitted, many video processing systems have been designed. Such systems encode video streams and transmit them over various kinds of networks or channels. In the ages of analog video systems, it was possible to evaluate the quality aspects of a video processing system by calculating the system's frequency response using test signals (for example, a collection of color bars and circles).

Digital video systems have almost fully replaced analog ones, and quality evaluation methods have changed. The performance of a digital video processing and transmission system can vary significantly and depends, amongst others, on the characteristics of the input video signal (e.g. amount of motion or spatial details), the settings used for encoding and transmission, and the channel fidelity or network performance.

Objective Video Quality

Objective video quality models are mathematical models that approximate results from subjective quality assessment, in which human observers are asked to rate the quality of a video. In this context, the term *model* may refer to a simple statistical model in which several independent variables (e.g. the packet loss rate on a network and the video coding parameters) are fit against results obtained in a subjective quality evaluation test using regression techniques. A model may also be a more complicated algorithm implemented in software or hardware.

Terminology

The terms *model* and *metric* are often used interchangeably in the field. However a metric has certain mathematical properties, which, by strict definition, do not apply to all video quality models.

The term "objective" relates to the fact that, in general, quality models are based on criteria that can be *measured* objectively – that is, free from human interpretation. They can be automatically evaluated by a computer program. Unlike a panel of human observers, an objective model should always deterministically output the same quality score for a given set of input parameters.

Objective quality models are sometimes also referred to as *instrumental (quality) models*, in order to emphasize their application as measurement instruments. Some authors suggest that the term "objective" is misleading, as it "implies that instrumental measurements bear objectivity, which they only do in case that they can be generalized."

Classification of Objective Video Quality Models

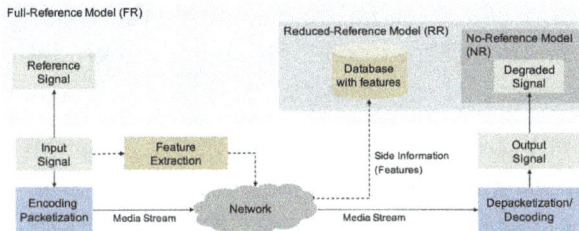

Classification of objective video quality models into Full-Reference, Reduced-Reference and No-Reference.

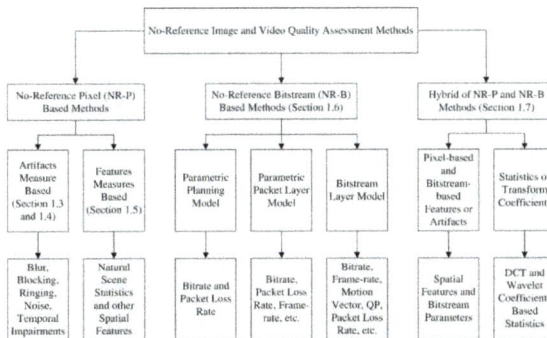

No-reference image and video quality assessment methods.

Objective models can be classified by the amount of information available about the original signal, the received signal, or whether there is a signal present at all:

- Full Reference Methods (FR): FR models compute the quality difference by comparing the original video signal against the received video signal. Typically, every pixel from the source is compared against the corresponding pixel at the received video, with no knowledge about the encoding or transmission process in between. More elaborate algorithms may choose to combine the pixel-based estimation with other approaches such as described below. FR models are usually the most accurate at the expense of higher computational effort. As they require availability of the original video before transmission or coding, they cannot be used in all situations (e.g., where the quality is measured from a client device).

- Reduced Reference Methods (RR): RR models extract some features of both videos and compare them to give a quality score. They are used when all the original video is not available, or when it would be practically impossible to do so, e.g. in a transmission with a limited bandwidth. This makes them more efficient than FR models at the expense of lower accuracy.

- No-Reference Methods (NR): NR models try to assess the quality of a distorted video without any reference to the original signal. Due to the absence of an original signal, they may be less accurate than FR or RR approaches, but are more efficient to compute.

  o Pixel-Based Methods (NR-P): Pixel-based models use a decoded representation of the signal and analyze the quality based on the pixel information. Some of these evaluate specific degradation types only, such as blurring or other coding artifacts.

  o Parametric/Bitstream Methods (NR-B): These models make use of features extracted from the transmission container and/or video bitstream, e.g. MPEG-TS packet headers, motion vectors and quantization parameters. They do not have access to the original signal and require no decoding of the video, which makes them more efficient. In contrast to NR-P models, they have no access to the final decoded signal. However, the picture quality predictions they deliver are not very accurate.

  o Hybrid Methods (Hybrid NR-P-B): Hybrid models combine parameters extracted from the bitstream with a decoded video signal. They are therefore a mix between NR-P and NR-B models.

Use of Picture Quality Models for Video Quality Estimation

Some models that are used for video quality assessment (such as PSNR or SSIM) are simply image quality models, whose output is calculated for every frame of a video sequence. This quality measure of every frame can then be recorded and pooled over time to assess the quality of an entire video sequence. While this method is easy to implement, it does not factor in certain kinds of degradations that develop over time, such as the moving artifacts caused by packet loss and its concealment. A video quality model that considers the temporal aspects of quality degradations, like VQM or the MOVIE Index, may be able to produce more accurate predictions of human-perceived quality.

Examples

No-reference Metrics

An overview of recent no-reference image quality models has been given in a journal paper by Shahid et al. As mentioned above, these can be used for video applications as well. No-reference, pixel-based quality models designed specifically for video are however rare, with Video-BLIINDS being one example. The Video Quality Experts Group has a dedicated working group on developing no-reference metrics (called NORM).

Simple Full-reference Metrics

The most traditional ways of evaluating quality of digital video processing system (e.g. a video codec) are FR-based. Among the oldest FR metrics are signal-to-noise ratio (SNR) and peak signal-to-noise ratio (PSNR), which are calculated between every frame of the original and the degraded video signal. PSNR is the most widely used objective image quality metric, and the average PSNR over all frames can be considered a video quality metric. PSNR is also used often during video codec development in order to optimize encoders. However, PSNR values do not correlate well with perceived picture quality due to the complex, highly non-linear behavior of the human visual system.

More Complex full- or Reduced-reference Metrics

With the success of digital video, a large number of more precise FR metrics have been developed. These metrics are inherently more complex than PSNR, and need more computational effort to calculate predictions of video quality. Among those metrics specifically developed for video are VQM and the MOVIE Index.

Based on the results of benchmarks by the Video Quality Experts Group (VQEG) (some in the course of the Multimedia Test Phase (2007–2008) and the HDTV Test Phase I (2009–2011)), some RR/FR metrics have been standardized in ITU-T as:

- ITU-T Rec. J.147 (FR), 2002 (includes VQM)
- ITU-T Rec. J.246 (RR), 2008
- ITU-T Rec. J.247 (FR), 2008
- ITU-T Rec. J.341 (FR), 2011
- ITU-T Rec. J.342 (RR), 2011

The popular Structural Similarity (SSIM) image quality metric is also often used for estimating video quality. Visual Information Fidelity (VIF) – also an image quality metric – is a core element of the Netflix Video Multimethod Assessment Fusion (VMAF), a tool that combines existing metrics to predict video quality.

Bitstream-based Metrics

Full or reduced-reference metrics still require access to the original video bitstream before transmission, or at least part of it. In practice, an original stream may not always be available for comparison, for example when measuring the quality from the user side. In other situations, a network operator may want to measure the quality of video streams passing through their network, without fully decoding them. For a more efficient estimation of video quality in such cases, parametric/bitstream-based metrics have also been standardized:

- ITU-T Rec. P.1201, 2012
- ITU-T Rec. P.1202, 2012
- ITU-T Rec. P.1203.1, 2016

Use in Practice

Few of these standards have found successful commercial application, including PEVQ and VQuad-HD. The Visual Information Fidelity (VIF) model, the Emmy-winning SSIM tool, the MOVIE Index and the older Tektronix PQA models are used by broadcast and post-production houses throughout the television and cinematic industries. VMAF is used by Netflix to tune their encoding and streaming algorithms, and to quality-control all streamed content.

Training and Performance Evaluation

Since objective video quality models are expected to predict results given by human observers, they are developed with the aid of subjective test results. During development of an objective model, its parameters should be trained so as to achieve the best correlation between the objectively predicted values and the subjective scores, often available as mean opinion scores (MOS).

The most widely used subjective test materials are in the public-domain and include still picture, motion picture, streaming video, high definition, 3-D (stereoscopic) and special-purposes picture quality related datasets. These so-called databases are created by various research laboratories around the world. Some of them have become de facto standards, including several public-domain subjective picture quality databases created and maintained by the Laboratory for Image and Video Engineering (LIVE) as well the Tampere Image Database 2008. A collection of databases can be found in the QUALINET Databases repository. The Consumer Digital Video Library (CDVL) hosts freely available video test sequences for model development.

In theory, a model can be trained on a set of data in such a way that it produces perfectly matching scores on that dataset. However, such a model will be over-trained and will therefore not perform well on new datasets. It is therefore advised to validate models against new data and use the resulting performance as a real indicator of the model's prediction accuracy.

To measure the performance of a model, some frequently used metrics are the linear correlation coefficient, Spearman's rank correlation coefficient, and the root mean square error (RMSE). Other metrics are the kappa coefficient and the outliers ratio. ITU-T Rec. P.1401 gives an overview of statistical procedures to evaluate and compare objective models.

Uses and Application of Objective Models

Objective video quality models can be used in various application areas. In video codec development, the performance of a codec is often evaluated in terms of PSNR or SSIM. For service providers, objective models can be used for monitoring a system. For example, an IPTV provider may choose to monitor their service quality by means of objective models, rather than asking users for their opinion, or waiting for customer complaints about bad video quality.

An objective model should only be used in the context that it was developed for. For example, a model that was developed using a particular video codec is not guaranteed to be accurate for another video codec. Similarly, a model trained on tests performed on a large TV screen should not be used for evaluating the quality of a video watched on a mobile phone.

Other Approaches

When estimating quality of a video codec, all the mentioned objective methods may require repeating post-encoding tests in order to determine the encoding parameters that satisfy a required level of visual quality, making them time consuming, complex and impractical for implementation in real commercial applications. There is ongoing research into developing novel objective evaluation methods which enable prediction of the perceived quality level of the encoded video before the actual encoding is performed.

Subjective Video Quality

The main goal of many objective video quality metrics is to automatically estimate the average user's (viewer's) opinion on the quality of a video processed by a system.

Procedures for subjective video quality measurements are described in ITU-R recommendation BT.500 and ITU-T recommendation P.910. In such tests, video sequences are shown to a group of viewers. The viewers' opinion is recorded and averaged into the Mean Opinion Score to evaluate the quality of each video sequence. However, the testing procedure may vary depending on what kind of system is tested.

Video Scaler

A video scaler is a system which is capable of converting video signals from one resolution to another. It increases or decreases the input resolution for video output at the specified ratio. Video scalers can be internal or external to the device. They are used in a wide range of applications such as broadcast, imaging, video effects and video surveillance.

Although all video displays have a built-in scaler for most varieties of inputs, the products are not designed for wide variety of formats and resolutions. On most digital devices, video scalers scale in both horizontal and vertical directions. If the resolution is increased from low to high, it is called upscaling, whereas if it is decreased from high to low, it is called downscaling. It typically accepts NTSC/PAL/SECAM signals and decodes them after which upscaling or downscaling is performed as needed. In other words, a video scaler first decodes the signal after which de-interlacing is done on it. Most of video scalers support bicubic, bilinear and polyphase scaling. A video scaler is considered essential if the input resolution is significantly different from that of the output, as without the video scaler, the images could be compromised or distorted.

Process

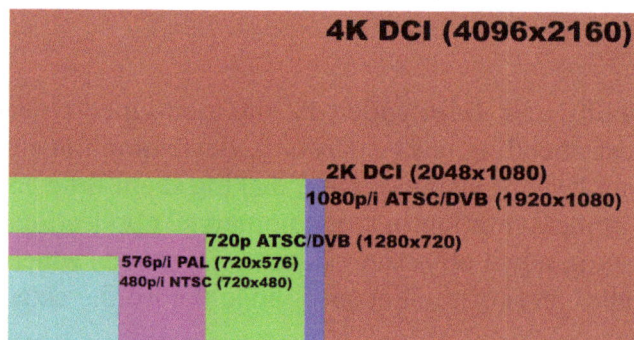

This is a comparison of several common video resolutions. The more pixels in an image the greater the possibility for finer detail and fidelity.

The "native resolution" of a display is how many physical pixels make up each row and column of the visible area on the display's output surface. There are many different video signals in use which are not the same resolution (neither are all of the displays), thus some form of resolution adaptation is required to properly frame a video signal to a display device. For example, within the United States, there are NTSC, ATSC, and VESA video standards each with several different resolution video formats. Multiple common resolutions are also used for high-definition television; 720p, 1080i, and 1080p.

While scaling a video signal does allow it to match the size of a particular display, the process can result in an increased number of visual artifacts in the signal, such as ringing and posterization.

Scaling by Television Channels

Television channels which air a mixture of 16:9 (or high definition) programming and 4:3 (or standard definition) programming may employ scaling and/or cropping in order to make the programming fill the entire screen, as opposed to pillarboxing the feed instead, in order to maintain consistency in format. Likewise, as opposed to "center-cropping", channels may downscale programming produced in 16:9 for broadcast on their 4:3 feeds through letterboxing—either as a full 16:9 letterbox, or a partial 14:9 letterbox—a technique used primarily by European broadcasters during the transition to digital terrestrial television. The Active Format Description standard is a system of variables defining various scaling, letterboxing, and pillarboxing states; broadcasting equipment and televisions can be configured to automatically switch to the appropriate state based on the AFD flag encoded in the content and the aspect ratio of the display.

When the U.S. cable network TNT introduced an HD feed in 2004, it controversially employed a stretching system known as FlexView (which was also offered to other broadcasters). FlexView used a nonlinear method to stretch more near the edges of the screen than in the center of it. The practice was imposed by the senior vice president of broadcast engineering at TNT, Clyde D. Smith, who argued that pillarboxing could cause burn-in on plasma televisions, some older HDTVs could not stretch 4:3 content automatically, the quality of stretching on some displays was poor, and also desired a more consistent viewing experience with no "jarring" transitions to 4:3 programming. Despite TNT's intentions, the system was frequently criticized by viewers of high definition channels, with some nicknaming the effect "Stretch-O-Vision".

In 2014, FXX faced similar criticism for its use of cropping and scaling on reruns of *The Simpsons* (which only started producing episodes in HD beginning in its 20th season), as its cropping method caused various visual gags to be lost. In February 2015, FXX announced that in response to these complaints, it would present these episodes in their original 4:3 aspect ratio on its video-on-demand service.

Benefits and Uses

There are many benefits associated with video scalers. They can output high quality, multiple-resolution VGA video signals, unlike a video switcher. They automatically scale any input signal received to match the resolution of the native display. They are also capable of switching between digital and analog sources.

Video scalers are typically used in consumer electronics such as televisions and AV devices.

Video File Format

Beginning from the home video era up to the most cutting-edge standards of today, video file formats have undergone some major changes. Different file formats do different things, and the right video format for a specific file isn't necessarily the right one for the others. Each file format has its own set of specifics.

With today's technology, the possibilities on personal computers and mobile devices seem to be endless, allowing us to create videos that grab the attention of an audience within seconds. Videos today are not just something that we watch, they are something that we engage in and something we become a part of. It is necessary to understand the different video file formats to ensure that the video is produced in the best format and quality for its intended purpose, hosting location, and audience.

Codecs

Codecs, deal with compressing your file. What is it, and why is it important? There are two kinds of compression: lossless and lossy. Most of the time, the quality can't be perceived by the human eye, but in some cases, it can make visuals look grainy, sounds flat and muffled, or make videos difficult to play.

In order to compress a video, your file must also have a corresponding codec. A codec is a software that compresses your video so it can be stored and played back. The most common codec includes h.264, which is often used for high-definition digital video and distribution of video content. It is also important to note the bit rate, which refers to the amount of data stored for each second of media that is played. The higher the bit rate, the less compression, which results in overall higher quality. However, be aware that the higher the bit rate, the larger the file size.

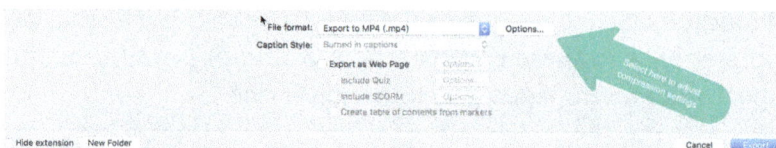

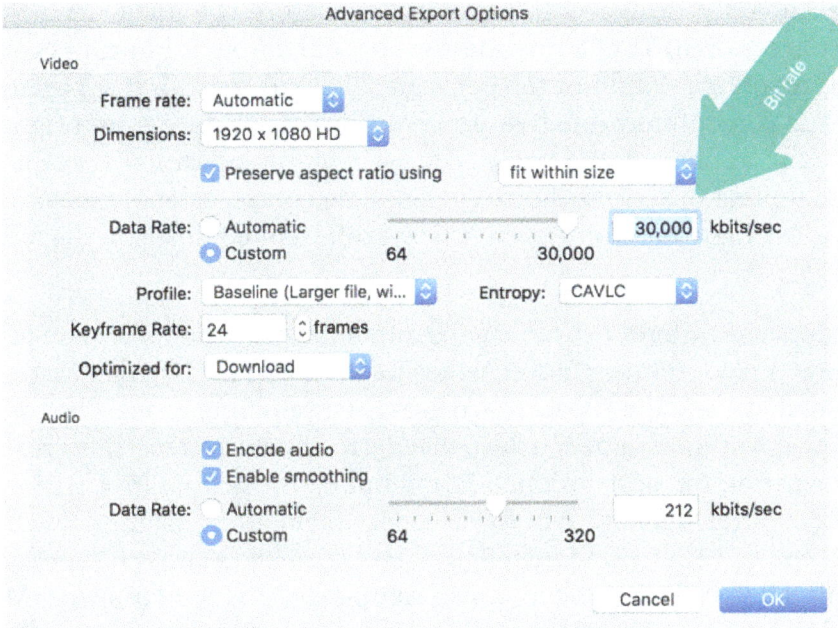

The "codec" can be dividing into 2 parts: encode and decode. The encoder performs the compression (encoding) function and the decoder performs the decompression (decoding) function. Some codecs include both of these components and some codecs only include one of them.

For example, when you rip a song from an audio CD to your computer, the Player uses the Windows Media Audio codec by default to compress the song into a compact WMA file. When you play that WMA file (or any WMA file that might be streamed from a website), the Player uses the Windows Media Audio codec to decompress the file so the music can be played through your speakers.

Applications

Video codecs are used in DVD players, Internet video, video on demand, digital cable, digital terrestrial television, videotelephony and a variety of other applications. In particular, they are widely used in applications that record or transmit video, which may not be feasible with the high data volumes and bandwidths of uncompressed video. For example, they are used in operating theaters to record surgical operations, in IP cameras in security systems, and in remotely operated underwater vehicles and unmanned aerial vehicles.

Video Codec Design

Video codecs seek to represent a fundamentally analog data set in a digital format. Because of the design of analog video signals, which represent luminance (luma) and color information (chrominance, chroma) separately, a common first step in image compression in codec design is to represent and store the image in a YCbCr color space. The conversion to YCbCr provides two benefits: first, it improves compressibility by providing decorrelation of the color signals; and second, it separates the luma signal, which is perceptually much more important, from the chroma signal, which is less perceptually important and which can be represented at lower resolution using chroma subsampling to achieve more efficient data compression. It is common to represent the ratios of information stored in these different channels in the following way Y:Cb:Cr.

Different codecs use different chroma subsampling ratios as appropriate to their compression needs. Video compression schemes for Web and DVD make use of a 4:2:1 color sampling pattern, and the DV standard uses 4:1:1 sampling ratios. Professional video codecs designed to function at much higher bitrates and to record a greater amount of color information for post-production manipulation sample in 3:1:1 (uncommon), 4:2:2 and 4:4:4 ratios. Examples of these codecs include Panasonic's DVCPRO50 and DVCPROHD codecs (4:2:2), and then Sony's HDCAM-SR (4:4:4) or Panasonic's HDD5 (4:2:2). Apple's Prores HQ 422 codec also samples in 4:2:2 color space. More codecs that sample in 4:4:4 patterns exist as well, but are less common, and tend to be used internally in post-production houses. It is also worth noting that video codecs can operate in RGB space as well. These codecs tend not to sample the red, green, and blue channels in different ratios, since there is less perceptual motivation for doing so—just the blue channel could be undersampled.

Some amount of spatial and temporal downsampling may also be used to reduce the raw data rate before the basic encoding process. The most popular such transform is the 8x8 discrete cosine transform (DCT). Codecs which make use of a wavelet transform are also entering the market, especially in camera workflows which involve dealing with RAW image formatting in motion sequences. The output of the transform is first quantized, then entropy encoding is applied to the quantized values. When a DCT has been used, the coefficients are typically scanned using a zig-zag scan order, and the entropy coding typically combines a number of consecutive zero-valued quantized coefficients with the value of the next non-zero quantized coefficient into a single symbol, and also has special ways of indicating when all of the remaining quantized coefficient values are equal to zero. The entropy coding method typically uses variable-length coding tables. Some encoders can compress the video in a multiple step process called *n-pass* encoding (e.g. 2-pass), which performs a slower but potentially better quality compression.

The decoding process consists of performing, to the extent possible, an inversion of each stage of the encoding process. The one stage that cannot be exactly inverted is the quantization stage. There, a best-effort approximation of inversion is performed. This part of the process is often called "inverse quantization" or "dequantization", although quantization is an inherently non-invertible process.

This process involves representing the video image as a set of macroblocks.

Video codec designs are usually standardized or eventually become standardized—i.e., specified precisely in a published document. However, only the decoding process need be standardized to enable interoperability. The encoding process is typically not specified at all in a standard, and implementers are free to design their encoder however they want, as long as the video can be decoded in the specified manner. For this reason, the quality of the video produced by decoding the results of different encoders that use the same video codec standard can vary dramatically from one encoder implementation to another.

Commonly used Video Codecs

A variety of video compression formats can be implemented on PCs and in consumer electronics equipment. It is therefore possible for multiple codecs to be available in the same product, avoiding the need to choose a single dominant video compression format for compatibility reasons.

Video in most of the publicly documented or standardized video compression formats can be created with multiple encoders made by different people. Many video codecs use common, standard video compression formats, which makes them compatible. For example, video created with a standard MPEG-4 Part 2 codec such as Xvid can be decoded (played back) using any other standard MPEG-4 Part 2 codec such as FFmpeg MPEG-4 or DivX Pro Codec, because they all use the same video format.

Some widely used software codecs are listed below, grouped by which video compression format they implement.

H.265/MPEG-H HEVC Codecs

- x265: A GPL-licensed implementation of the H.265 video standard. x265 is only an encoder.

H.264/MPEG-4 AVC Codecs

- X264: A GPL-licensed implementation of the H.264 video standard. x264 is only an encoder.
- Nero Digital: Commercial MPEG-4 ASP and AVC codecs developed by Nero AG.
- QuickTime H.264: H.264 implementation released by Apple.
- DivX Pro Codec: An H.264 decoder and encoder was added in version 7.

H.263/MPEG-4 Part 2 Codecs

- DivX Pro Codec: A proprietary MPEG-4 ASP codec made by DivX, Inc.
- Xvid: Free/open-source implementation of MPEG-4 ASP, originally based on the OpenDivX project.

- FFmpeg MPEG-4: Included in the open-source libavcodec codec library, which is used by default for decoding or encoding in many open-source video players, frame-works, editors and encoding tools such as MPlayer, VLC, ffdshow or GStreamer. Compatible with other standard MPEG-4 codecs like Xvid or DivX Pro Codec.

- 3ivx: A commercial MPEG-4 codec created by 3ivx Technologies.

H.262/MPEG-2 Codecs

- x262: A GPL-licensed implementation of the H.262 video standard. x262 is only an encoder.

Microsoft Codecs

- WMV (Windows Media Video): Microsoft's family of proprietary video codec designs including WMV 7, WMV 8, and WMV 9. The latest generation of WMV is standardized by SMPTE as the VC-1 standard.

- MS MPEG-4v3: A proprietary and not MPEG-4 compliant video codec created by Microsoft. Released as a part of Windows Media Tools 4. A hacked version of Microsoft's MPEG-4v3 codec became known as DivX ;-)

Google (On2) Codecs

- VP6, VP6-E, VP6-S, VP7, VP8, VP9: Proprietary high definition video compression formats and codecs developed by On2 Technologies used in platforms such as Adobe Flash Player 8 and above, Adobe Flash Lite, Java FX and other mobile and desktop video platforms. Supports resolution up to 720p and 1080p. VP9 supports resolutions up to 2160p. VP8 and VP9 have been available under the New BSD License by Google with source code available as the libvpx VP8/VP9 codec SDK.

- libtheora: A reference implementation of the Theora video compression format developed by the Xiph.org Foundation, based upon On2 Technologies' VP3 codec, and christened by On2 as the successor in VP3's lineage. Theora is targeted at competing with MPEG-4 video and similar lower-bitrate video compression schemes.

Lossless Codecs

Other Codecs

- Apple ProRes: Is a lossy video compression format developed by Apple Inc.

- Schrödinger and dirac-research: implementations of the Dirac compression format developed by BBC Research at the BBC. Dirac provides video compression from web video up to ultra HD and beyond.

- DNxHD codec: a lossy high-definition video production codec developed by Avid Technology. It is an implementation of VC-3.

- Sorenson 3: A video compression format and codec that is popularly used by Apple's QuickTime, sharing many features with H.264. Many movie trailers found on the web use this compression format.

- Sorenson Spark: A codec and compression format that was licensed to Macromedia for use in its Flash Video starting with Flash Player 6. It is considered as an incomplete implementation of the H.263 standard.

- RealVideo: Developed by RealNetworks. A popular compression format and codec technology a few years ago, now fading in importance for a variety of reasons.

- Cinepak: A very early codec used by Apple's QuickTime.

- Indeo, an older video compression format and codec initially developed by Intel.

All of the codecs above have their qualities and drawbacks. Comparisons are frequently published. The trade-off between compression power, speed, and fidelity (including artifacts) is usually considered the most important figure of technical merit.

Codec Packs

Online video material is encoded by a variety of codecs, and this has led to the availability of codec packs — a pre-assembled set of commonly used codecs combined with an installer available as a software package for PCs.

Containers

In addition to a codec, each video file has a container. The container is like a box that contains your video, audio, and metadata (vital data such as captions, SEO, and vital information that pieces the video together for playback). It can also be called a file extension since they are often seen as file names, such as AVI, MOV, or MP4.

Some Video File Format Types

AVI (Audio Video Interleave)

Developed by Microsoft and introduced to the public in November 1992 as part of its Video for Windows technology, the AVI format is one of the oldest video formats. It is so universally accepted that many people consider it the de facto standard for storing video and audio information on the computer. Due to it's simple architecture, AVI files are able to run on a number of different systems like Windows, Macintosh, Linux; is also supported by popular web browsers. AVI files stores data that can be encoded in a number of different codec's, although most commonly with M-JPEG or DivX codecs. This means that all AVI files, while they may look similar on the outside, differ substantially from one another on the inside.

FLV (Flash Video Format)

FLV files are videos that are encoded by Adobe Flash software, usually with codecs following the Sorenson Spark or VP6 video compression formats. They can be played via the Adobe Flash Player, web browser plugins or one of several third party programs. Since virtually everyone has the player installed on their browsers, it has become the most common online video viewing platform used on the Web today. As almost all video sharing sites such as Youtube stream videos in Flash, practically all browsers support and are compatible with the Flash Video format and can play the video with ease. In addition to being an online video viewing format, the Flash Video format is also what many video-sharing sites convert videos to, from formats that were uploaded by their users in something other than Flash. This is because videos in the FLV format remain in high quality even after compression to a smaller file size, which means that the videos on the Web load quickly and won't spend a lot of time using up bandwidth. Some notable users of the Flash Video are Youtube, Yahoo! Video, VEVO, Hulu and Myspace among many others.

WMV (Windows Media Video)

Developed by Microsoft, WMV was originally designed for web streaming applications, as a competitor to RealVideo, but it can now cater to more specialized content. WMV files are the tiniest video files over the Web, as their file size decreases significantly after compression, which results in poor video quality. However, one advantage of this small file size is that it is probably the only video file format that allows users to upload and share their videos through the e-mail system. Being a Microsoft software, the Windows Media Player is the main application that is used to play WMV files on all Microsoft's Windows operating systems, but there are also WMV players available for free for the Macintosh operating system.

MOV (Apple QuickTime Movie)

Developed by Apple. Inc, the QuickTime file format is a popular type of video sharing and viewing format amongst Macintosh users, and is often used on the Web, and for saving movie and video files. In recent years, Apple came up with a newer version called QuickTime X, currently available on Mac OS X Snow Leopard, Lion and Mountain Lion. MOV files are most commonly opened via the Apple QuickTime Player for the Macintosh Operating System. However, MOV files are not just limited to being played on Apple computers, as there is a free version of the QuickTime Player available for the Windows Operating System among many other players. Considered one of the best looking file formats, MOV files are of high quality and are usually big in file size.

MP4 (Moving Pictures Expert Group 4)

MP4 is an abbreviated term for MPEG-4 Part 14, a standard developed by the Motion Pictures Expert Group who was responsible for setting industry standards regarding digital audio and video, and is commonly used for sharing video files on the Web. First introduced in 1998, the MPEG-4 video format uses separate compression for audio and video tracks; video is compressed with MPEG-4 or H.264 video encoding; and audio is compressed using AAC compression. The MP4 file format is also another great file sharing format for the Web, MP4 file sizes are relatively small but the quality remains high even after compression. MP4 standard is also becoming more popular than FLV for online video sharing, as it compatible with both online and mobile browsers and also supported by the new HTML5.

Video Coding Format

Videos take up a lot of space—just how much varies widely depending on the video format, the resolution, and the number of frames per second that was chosen when the video was created.

Uncompressed 1080 HD video footage takes up about 10.5 GB of space per minute of video. If you use a smartphone to shoot your video, 1080p footage takes up 130 MB per minute of footage, while 4K video takes up 375 MB of space for each minute of film.

Because videos take up so much space, and because bandwidth is limited, video files are almost always compressed before being put on the web (or downloaded from the web). Compression involves packing the file's information into a smaller space.

A video coding format (or sometimes video compression format) is a content representation format for storage or transmission of digital video content (such as in a data file or bitstream). Examples of video coding formats include MPEG-2 Part 2, MPEG-4 Part 2, H.264 (MPEG-4 Part 10), HEVC, Theora, RealVideo RV40, VP9, and AV1. A specific software or hardware implementation capable of video compression and/or decompression to/from a specific video coding format is called a video codec; an example of a video codec is Xvid, which is one of several different codecs which implements encoding and decoding videos in the MPEG-4 Part 2 video coding format in software.

Some video coding formats are documented by a detailed technical specification document known as a video coding specification. Some such specifications are written and approved by standardization organizations as technical standards, and are thus known as a video coding standard. The term 'standard' is also sometimes used for de facto standards as well as formal standards.

Video content encoded using a particular video coding format is normally bundled with an audio stream (encoded using an audio coding format) inside a multimedia container format such as AVI, MP4, FLV, RealMedia, or Matroska. As such, the user normally doesn't have a H.264 file, but instead has a .mp4 video file, which is an MP4 container containing H.264-encoded video, normally alongside AAC-encoded audio. Multimedia container formats can contain any one of a number of different video coding formats; for example the MP4 container format can contain video in either the MPEG-2 Part 2 or the H.264 video coding format, among others. Another example is the initial specification for the file type WebM, which specified the container format (Matroska), but also exactly which video (VP8) and audio (Vorbis) compression format is used inside the Matroska container, even though the Matroska container format itself is capable of containing other video coding formats (VP9 video and Opus audio support was later added to the WebM specification).

Distinction between "Format" and "Codec"

Although video coding formats such as H.264 are sometimes referred to as codecs, there is a clear conceptual difference between a specification and its implementations. Video coding formats are described in specifications, and software or hardware to encode/decode data in a given video coding format from/to uncompressed video are implementations of those specifications. As an analogy, the video coding format H.264 (specification) is to the codec OpenH264 (specific implementation) what the C Programming Language (specification) is to the compiler GCC (specific implementation). Note that for each specification (e.g. H.264), there can be many codecs implementing that specification (e.g. x264, OpenH264, H.264/MPEG-4 AVC products and implementations).

This distinction is not consistently reflected terminologically in the literature. The H.264 specification calls H.261, H.262, H.263, and H.264 video coding standards and does not contain the word codec. The Alliance for Open Media clearly distinguishes between the AV1 video coding format and the accompanying codec they are developing, but calls the video coding format itself a video codec specification. The VP9 specification calls the video coding format VP9 itself a codec.

As an example of conflation, Chromium's and Mozilla's pages listing their video format support both call video coding formats such as H.264 codecs. As another example, in Cisco's announcement of a free-as-in-beer video codec, the press release refers to the H.264 video coding format as a "codec" ("choice of a common video codec"), but calls Cisco's implementation of a H.264 encoder/decoder a "codec" shortly thereafter ("open-source our H.264 codec").

A video coding format does not dictate all algorithms used by a codec implementing the format. For example, a large part of how video compression typically works is by finding similarities between video frames (block-matching), and then achieving compression by copying previously-coded similar subimages (e.g., macroblocks) and adding small differences when necessary. Finding optimal combinations of such predictors and differences is an NP-hard problem, meaning that it is practically impossible to find an optimal solution. While the video coding format must support such compression across frames in the bitstream format, by not needlessly mandating specific algorithms for finding such block-matches and other encoding steps, the codecs implementing the video coding specification have some freedom to optimize and innovate in their choice of algorithms. For example, section 0.5 of the H.264 specification says that encoding algorithms are not part of the specification. Free choice of algorithm also allows different space–time complexity trade-offs for the same video coding format, so a live feed can use a fast but space-inefficient algorithm, while a one-time DVD encoding for later mass production can trade long encoding-time for space-efficient encoding.

Lossless, Lossy, and Uncompressed Video Coding Formats

Consumer video is generally compressed using lossy video codecs, since that results in significantly smaller files than lossless compression. While there are video coding formats designed explicitly for either lossy or lossless compression, some video coding formats such as Dirac and H.264 support both.

Uncompressed video formats, such as *Clean HDMI*, is a form of lossless video used in some circumstances such as when sending video to a display over a HDMI connection. Some high-end cameras can also capture video directly in this format.

Intra-frame Video Coding Formats

Interframe compression complicates editing of an encoded video sequence. One subclass of relatively simple video coding formats are the intra-frame video formats, such as DV, in which each frame of the video stream is compressed independently without referring to other frames in the stream, and no attempt is made to take advantage of correlations between successive pictures over time for better compression. One example is Motion JPEG, which is simply a sequence of individually JPEG-compressed images. This approach is quick and simple, at the expense the encoded video being much larger than a video coding format supporting Inter frame coding.

Because interframe compression copies data from one frame to another, if the original frame is simply cut out (or lost in transmission), the following frames cannot be reconstructed properly. Making 'cuts' in intraframe-compressed video while video editing is almost as easy as editing uncompressed video: one finds the beginning and ending of each frame, and simply copies bit-for-bit each frame that one wants to keep, and discards the frames one doesn't want. Another difference between intraframe and interframe compression is that, with intraframe systems, each frame uses a similar amount of data. In most interframe systems, certain frames (such as "I frames" in MPEG-2) aren't allowed to copy data from other frames, so they require much more data than other frames nearby.

It is possible to build a computer-based video editor that spots problems caused when I frames are edited out while other frames need them. This has allowed newer formats like HDV to be used for editing. However, this process demands a lot more computing power than editing intraframe compressed video with the same picture quality.

Profiles and Levels

A video coding format can define optional restrictions to encoded video, called profiles and levels. It is possible to have a decoder which only supports decoding a subset of profiles and levels of a given video format, for example to make the decoder program/hardware smaller, simpler, or faster.

A *profile* restricts which encoding techniques are allowed. For example, the H.264 format includes the profiles *baseline*, *main* and *high* (and others). While P-slices (which can be predicted based on preceding slices) are supported in all profiles, B-slices (which can be predicted based on both preceding and following slices) are supported in the *main* and *high* profiles but not in *baseline*.

A *level* is a restriction on parameters such as maximum resolution and data rates.

Internet Video Coding

Video encoding is the process of converting digital video files from one standard digital video format into another. The purpose of this is for compatibility and efficiency with a desired set of applications and hardware such as for DVD/Blu-ray, mobile, video streaming or general video editing. The encoding process transforms the video and audio data in the file and then does compression according to the specifications of the encoding standard chosen.

Video encoding is the process of changing a digital video's format from one standard into another generally for the purpose of compatibility. This is because digital video can exist in different formats with different variables such as containers like .mp4, .flv, .avi and .wmv, and can have different codecs (which facilitate the compression/decompression) and, hence, different qualities meant for different applications.

Video encoding is therefore simply the process of preparing a video for output, which greatly varies depending on the intent and use. For example, videos meant for DVD have to be in MPEG-2 format, whereas those for Blu-ray use H.264/MPEG-4 AVC, which YouTube also uses currently after it moved from the FLV format.

Internet Video Coding (IVC) has been developed in MPEG by combining well-known existing technology elements and new coding tools with royalty-free declarations. In June 2015, IVC project was approved as ISO/IEC 14496-33 (MPEG- 4 Internet Video Coding). It is believed that this standard can be highly beneficial for video services in the Internet domain.

Experimental results show that IVC's compression performance is approximately equal to that of the AVC High Profile for typical operational settings, both for streaming and low-delay applications, and is better than WVC and VCB.

Intra-frame Coding

The term intra frame coding refers to the fact that the various lossless and lossy compression techniques are performed relative to information that is contained only within the current frame, and not relative to any other frame in the video sequence. In other words, no temporal processing is performed outside of the current picture or frame.

Figure shows a block diagram of a basic video encoder for intra frames only. It turns out that this block diagram is very similar to that of a JPEG still image video encoder, with only slight implementation detail differences.

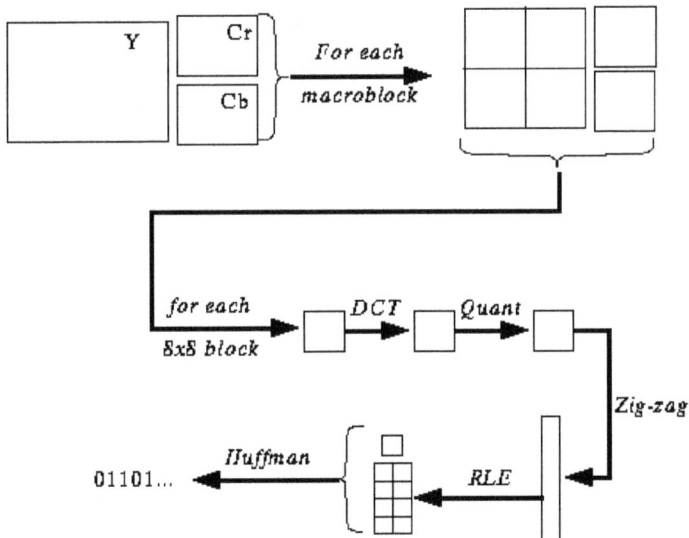

The basic processing blocks shown are the video filter, discrete cosine transform, DCT coefficient quantizer, and run-length amplitude/variable length coder. These blocks are described individually in the sections below or have already been described in JPEG Compression.

This is a basic Intra Frame Coding Scheme is as follows:

- Macroblocks are 16x16 pixel areas on Y plane of original image.

  A macroblock usually consists of 4 Y blocks, 1 Cr block, and 1 Cb block.

  In the example HDTV data rate calculation shown previously, the pixels were represented as 8-bit values for each of the primary colors , red, green, and blue. It turns out that while this may be good for high performance computer generated graphics, it is wasteful in most video compression applications. Research into the Human Visual System (HVS) has shown that the eye is most sensitive to changes in luminance, and less sensitive to variations in chrominance. Since absolute compression is the name of the game, it makes sense that MPEG should operate on a color space that can effectively take advantage of the eye's different sensitivity to luminance and chrominance information. As such, H/261 (and MPEG) uses the YCbCr color space to represent the data values instead of RGB, where Y is the luminance signal, Cb is the blue color difference signal, and Cr is the red color difference signal.

  A macroblock can be represented in several different manners when referring to the YCbCr color space. below shows 3 formats known as 4:4:4, 4:2:2, and 4:2:0 video. 4:4:4 is full bandwidth YCbCr video, and each macroblock consists of 4 Y blocks, 4 Cb blocks, and 4 Cr blocks. Being full bandwidth, this format contains as much information as the data would if it were in the RGB color space. 4:2:2 contains half as much chrominance information as 4:4:4, and 4:2:0 contains one quarter of the chrominance information. Although MPEG-2 has provisions to handle the higher chrominance formats for professional applications, most consumer level products will use the normal 4:2:0 mode.

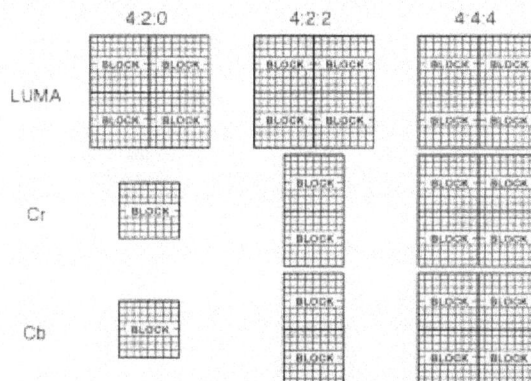

Macroblock Video Formats

Because of the efficient manner of luminance and chrominance representation, the 4:2:0 representation allows an immediate data reduction from 12 blocks/macroblock to 6 blocks/macroblock, or 2:1 compared to full bandwidth representations such as 4:4:4 or RGB. To generate this format without generating color aliases or artifacts requires that the chrominance signals be filtered.

The Macroblock is coded as follows:

Addr	Type	Quant	Vector	CBP	b0	b1	...	b5

- o   Many macroblocks will be exact matches (or close enough). So send address of each block in image -> *Addr*

- o   Sometimes no good match can be found, so send INTRA block -> *Type*

- o   Will want to vary the quantization to fine tune compression, so send quantization value -> *Quant*

- o   Motion vector -> *vector*

- o   Some blocks in macroblock will match well, others match poorly. So send bitmask indicating which blocks are present (Coded Block Pattern, or *CBP*).

- o   Send the blocks (4 Y, 1 Cr, 1 Cb) as in JPEG.

- •   Quantization is by constant value for all DCT coefficients (i.e., no quantization table as in JPEG).

Coding Process

Data is usually read from a video camera or a video card in the YCbCr data format (often informally called YUV for brevity). The coding process varies greatly depending on which type of encoder is used (e.g., JPEG or H.264), but the most common steps usually include: partitioning into macroblocks, transformation (e.g., using a DCT or wavelet), quantization and entropy encoding.

Multiview Video Coding

The Multiview Video Coding amendment (MVC) of the H.264/AVC standard provides view scalability at the bitstream level. This allows the efficient transmission of multiview video (e.g., video with 2 views suitable for viewing on a stereo display) in an efficient and backward compatible way. This is illustrated in figure for a bitstream with 2 views. A legacy H.264/AVC decoder decodes only one (the so-called base view) of the two views that are included in the multiview bitstream. The reconstructed video sequence can be displayed on a conventional 2d display. On the contrary, a stereo decoder is capable of decoding both views and the decoded video sequences (one for the left and one for the right eye) are suitable for 3d displays.

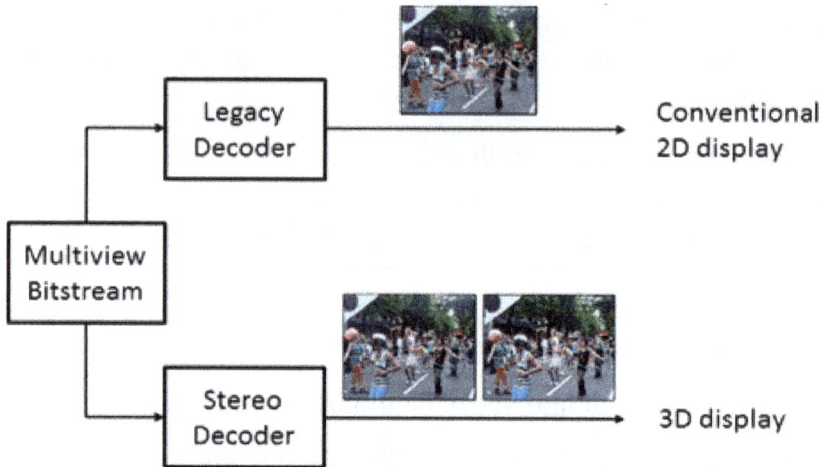

The desire for an efficient representation of multiview video originates from the growing interest in 3D video. One of the premium application areas of MVC are 3D-Blu-Ray players. Conventional Blu-Ray players can decode one of the two views that are stored on a 3D-Blue-Ray disc and ignore the data on the disc for the other view. 3D-capable Blu-Ray players, however, can decode both views, which can then be displayed on a 3D display.

A simple approach for multiview video coding is to code all of the multiple views with a legacy 2d video coding standard. Although such a concept provides the functionality required for multiview video coding, it does not use the dependencies between views and thus does not provide a suitable coding efficiency. Besides the exploitation of temporal dependencies, it is required to also exploit the dependencies between the different views for an efficient coding. On the other side, for an adoption of a multiview coding standard by the industry it was desirable to modify as less aspects of the original H.264/AVC design as possible.

In collaboration with the 3D Coding Group, the Image and Video Coding Group developed a simple, but very efficient concept for extending the H.264/AVC standard for multiview video coding. The base view is coded using the unmodified H.264/AVC syntax, so that legacy decoders can always decode the base view. As another important aspect, the MVC extension does not include any changes of the basic decoding process. Only the high-level syntax was modified. For efficiently exploiting the inter-view dependencies, the concept of multi-frame motion-compensated prediction, which was developed by the Image and Video Coding Group, was extended. In conventional 3d video coding, multiple previously decoded picture can be inserted into so-called reference picture lists. When coding a block using motion-compensated prediction (MCP), a reference picture index, which signals the used reference picture inside a reference picture list, is transmitted in addition to a motion vector. For dependent views, not only previously decoded pictures of the same view can be inserted into the reference picture lists, but also already decoded pictures of other views for the same access unit (time instant). Here the decoded picture of the base view of the same access unit is inserted at position 1 of the reference picture list. Thus, when a reference index equal to 1 is transmitted for a block, it signals that disparity-compensated prediction (DCP) using the base view picture of the same time instant is used. If any other reference index is transmitted, conventional motion-compensated prediction (MCP) using a previously decoded picture of the same view is used.

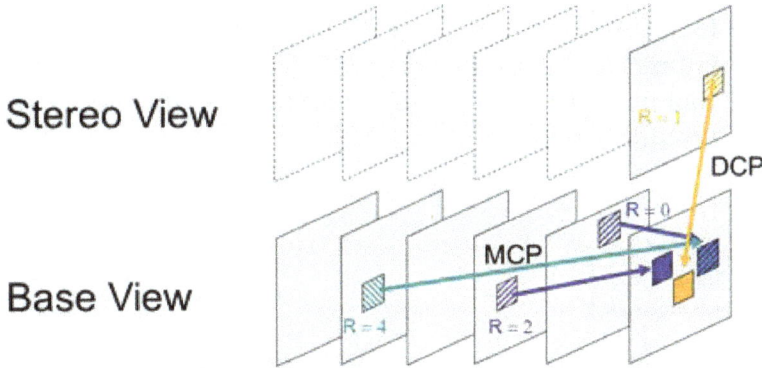

The investigations for multiview video coding also showed that the disparity-compensated prediction, and thus the entire multiview video coding approach, is particularly efficient if it is combined with the concept of hierarchical prediction structures, which have been previously developed by the Image and Video Coding Group.

The joint proposal of Image and Video Coding Group and the 3D Coding Group has been chosen as the starting point of the MVC development and has been adopted with some additional high-level signaling mechanisms as the final MVC standard.

Open Source Support Mostly Missing

As of April 2015 there is still no free and open source software that supports software decoding of the MVC video compression standard. So popular open source H.264 and HEVC (H.265) decoders such as those used in the FFmpeg and Libav libraries simply ignore the additional information for the second view and thus do not show the second view for stereoscopic views. In most cases the reason for this support not being added is that MVC was not considered when the initial core H.264 and HEVC decoders code was written so it was coded in one large chunk, and later amendment would as such often mean a lot of prerequisite code refactoring work and large changes its current architecture, with major work in untangling and reordering some code, and splitting different functions in existing decoder code into smaller chunks for simpler handling to in turn then make amendments such as MVC easier to add.

Some proof-of-concept work has however been done downstream in the past but never made it upstream into official releases of FFmpeg or Libav.

On March 8, 2016, the situation improved. Version 0.68 of the DirectShow Media Splitter and Decoders Collection LAV Filters was released by developer "Nevcairiel" (who also works for Media Player Classic - Home Cinema (MPC-HC)) with support of H.264 MVC 3D demuxing and decoding. With the aid of this release and FRIM written by a programmer named "videohelp3d" it is possible to write an AviSynth script to pre process a H.264 MVC 3D video clip which can then be opened by free 3D video player Bino and then shown as red - cyan anaglyph video for example.

The usage of the FRIM AviSynth plugin (FRIMSource) is described on "videohelp3d" home page. LAV Filters can be used to get audio from H.264 MVC 3D video clip. The developer posted that in a future release of it might be possible that LAV Video renders the video as Side-by-Side directly.

Digital Video

Digital video (DV) is video that is captured and stored in a digital format as ones and zeros, rather than a series of still pictures captured in film. Digital, versus analog, signals are used. Information is processed and stored as a sequence of digital data for easy manipulation by computers, but the video is still presented to the viewer through a screen in analog form.

Digital video is composed of a series of orthogonal bitmap (BMP) images displayed in constant rapid succession with common frequencies of 15, 24, 30 and 60 frames per second (FPS); the more frames the DV has, the more movement details are captured or displayed.

As a point of reference, good quality movies and videos are recorded and viewed at 60 FPS, while super slow motion videos are taken with high-speed photography equipment at more than 1,000 FPS and then viewed at standard rates. Each orthogonal BMP image or frame in the DV includes a raster of pixels with width and height expressed in number of pixels, known as resolution. The higher the captured video's resolution, the higher its clarity and quality.

Because of digital manipulation, a video can be upscaled, or captured in low resolution and displayed in higher resolution with obvious losses in perceived and numerical quality. However, a high resolution video can be successfully downscaled without perceived quality loss, even though the images are perceivably smaller and, thus, of a lower quality on a high resolution screen.

Digital video comprises a series of orthogonal bitmap digital images displayed in rapid succession at a constant rate. In the context of video these images are called frames.We measure the rate at which frames are displayed in frames per second (FPS). Since every frame is an orthogonal bitmap digital image it comprises a raster of pixels. If it has a width of W pixels and a height of H pixels we say that the frame size is WxH. Pixels have only one property, their color. The color of a pixel is represented by a fixed number of bits. The more bits the more subtle variations of colors can be reproduced. This is called the color depth (CD) of the video.

An example video can have a duration (T) of 1 hour (3600sec), a frame size of 640x480 (WxH) at a color depth of 24bits and a frame rate of 25fps. This example video has the following properties:

- pixels per frame = 640 * 480 = 307,200

- bits per frame = 307,200 * 24 = 7,372,800 = 7.37Mbits

- bit rate (BR) = 7.37 * 25 = 184.25Mbits/sec

- video size (VS) = 184Mbits/sec * 3600sec = 662,400Mbits = 82,800Mbytes = 82.8Gbytes

The most important properties are bit rate and video size. The formulas relating those two with all other properties are:

$$BR = W * H * CD * FPS$$

$$VS = BR * T = W * H * CD * FPS * T$$

(units are: BR in bit/s, W and H in pixels, CD in bits, VS in bits, T in seconds)

while some secondary formulas are:

$$pixels\_per\_frame = W * H$$

$$pixels\_per\_second = W * H * FPS$$

$$bits\_per\_frame = W * H * CD$$

Interlacing

In interlaced video each *frame* is composed of two *halves of an image*. The first half contains only the odd-numbered lines of a full frame. The second half contains only the even-numbered lines. Those halves are referred to individually as *fields*. Two consecutive fields compose a full frame. If an interlaced video has a frame rate of 15 frames per second the field rate is 30 fields per second. All the properties and formulas discussed here apply equally to interlaced video but one should be careful not to confuse the fields per second rate with the frames per second rate.

Properties of Compressed Video

Because of the relatively high bit rate of uncompressed video, video compression is extensively used. In the case of compressed video each frame requires a small percentage of the original bits. Assuming a compression algorithm that shrinks the input data by a factor of CF, the bit rate and video size would equal to:

$$BR = W * H * CD * FPS/CF$$

$$VS = BR * T/CF$$

Note that it is not necessary that all frames are equally compressed by a factor of CF. In practice they are not, so CF is the *average* factor of compression for *all* the frames taken together.

The above equation for the bit rate can be rewritten by combining the compression factor and the color depth like this:

$$BR = W * H * (CD/CF) * FPS$$

The value (CD/CF) represents the average bits per pixel (BPP). As an example, if we have a color depth of 12bits/pixel and an algorithm that compresses at 40x, then BPP equals 0.3 (12/40). So in the case of compressed video the formula for bit rate is:

$$BR = W * H * BPP * FPS$$

The same formula is valid for uncompressed video because in that case one can assume that the "compression" factor is 1 and that the average bits per pixel equal the color depth.

Bit Rate and BPP

By its definition, bit rate is a measure of the rate of information content of the digital video stream. In the case of uncompressed video, bit rate corresponds directly to the quality of the video. (Bit rate is proportional to every property that affects the video quality.) Bit rate is an important property when transmitting video because the transmission link must be capable of supporting that bit rate. Bit rate is also important when dealing with the storage of video because, the video size is proportional to the bit rate and the duration. Bit rate of uncompressed video is too high for most practical applications. Video compression is used to greatly reduce the bit rate. BPP is a measure of the efficiency of compression. A true-color video with no compression at all may have a BPP of 24 bits/pixel. Chroma subsampling can reduce the BPP to 16 or 12 bits/pixel. Applying jpeg compression on every frame can reduce the BPP to 8 or even 1 bits/pixel. Applying video compression algorithms like MPEG1, MPEG2 or MPEG4 allows for fractional BPP values.

Constant Bit Rate versus Variable Bit Rate

BPP represents the *average* bits per pixel. There are compression algorithms that keep the BPP almost constant throughout the entire duration of the video. In this case, we also get video output with a constant bit rate (CBR). This CBR video is suitable for real-time, non-buffered, fixed bandwidth video streaming (e.g. in videoconferencing).

As not all frames can be compressed at the same level, because quality is more severely impacted for scenes of high complexity, some algorithms try to constantly adjust the BPP. They keep it high while compressing complex scenes and low for less demanding scenes. This way, one gets the best quality at the smallest average bit rate (and the smallest file size, accordingly). When using this method, the bit rate is variable because it tracks the variations of the BPP.

Technical Overview

Standard film stocks such as 16 mm and 35 mm record at 24 frames per second. For video, there are two frame rate standards: NTSC, which shoot at 30/1.001 (about 29.97) frames per second or 59.94 fields per second, and PAL, 25 frames per second or 50 fields per second. Digital video cameras come in two different image capture formats: interlaced and deinterlaced/progressive scan. Interlaced cameras record the image in alternating sets of lines: the odd-numbered lines are scanned, and then the even-numbered lines are scanned, then the odd-numbered lines are scanned again, and so on. One set of odd or even lines is referred to as a "field", and a consecutive pairing of two fields of opposite parity is called a *frame*. Deinterlaced cameras records each frame as distinct, with all scan lines being captured at the same moment in time. Thus, interlaced video captures samples the scene motion twice as often as progressive video does, for the same number of frames per second. Progressive-scan camcorders generally produce a slightly sharper image. However, motion may not be as smooth as interlaced video which uses 50 or 59.94 fields per second, particularly if they employ the 24 frames per second standard of film.

Digital video can be copied with no degradation in quality. No matter how many generations of a digital source is copied, it will still be as clear as the original first generation of digital footage. However a change in parameters like frame size as well as a change of the digital format can decrease the quality of the video due to new calculations that have to be made. Digital video can be manipulated and edited to follow an order or sequence on an NLE, or non-linear editing workstation, a computer-based device intended to edit video and audio. More and more, videos are edited on readily available, increasingly affordable consumer-grade computer hardware and software. However, such editing systems require ample disk space for video footage. The many video formats and parameters to be set make it quite impossible to come up with a specific number for how many minutes need how much time.

Digital video has a significantly lower cost than 35 mm film. In comparison to the high cost of film stock, the tape stock (or other electronic media used for digital video recording, such as flash memory or hard disk drive) used for recording digital video is very inexpensive. Digital video also allows footage to be viewed on location without the expensive chemical processing required by film. Also physical deliveries of tapes and broadcasts do not apply anymore. Digital television (including higher quality HDTV) started to spread in most developed countries in early 2000s. Digital video is also used in modern mobile phones and video conferencing systems. Digital video is also used for Internet distribution of media, including streaming video and peer-to-peer movie distribution. However even within Europe are lots of TV-Stations not broadcasting in HD, due to restricted budgets for new equipment for processing HD.

Many types of video compression exist for serving digital video over the internet and on optical disks. The file sizes of digital video used for professional editing are generally not practical for these purposes, and the video requires further compression with codecs such as Sorenson, H.264 and more recently Apple ProRes especially for HD. Probably the most widely used formats for delivering video over the internet are MPEG4, Quicktime, Flash and Windows Media, while MPEG2 is used almost exclusively for DVDs, providing an exceptional image in minimal size but resulting in a high level of CPU consumption to decompress.

As of 2011, the highest resolution demonstrated for digital video generation is 35 megapixels (8192 x 4320). The highest speed is attained in industrial and scientific high speed cameras that are capable of filming 1024x1024 video at up to 1 million frames per second for brief periods of recording.

Interfaces and Cables

Many interfaces have been designed specifically to handle the requirements of uncompressed digital video (from roughly 400 Mbit/s to 10 Gbit/s):

- High-Definition Multimedia Interface
- Digital Visual Interface
- Serial Digital Interface
- DisplayPort
- Digital component video
- Unified Display Interface
- FireWire
- USB

The following interface has been designed for carrying MPEG-Transport compressed video:

- DVB-ASI

Compressed video is also carried using UDP-IP over Ethernet. Two approaches exist for this:

- Using RTP as a wrapper for video packets

- 1-7 MPEG Transport Packets are placed directly in the UDP packet

Alternative method of carrying video over IP include:

- Network Device Interface

Storage Formats

Encoding

All current formats, which are listed below, are PCM based.

- CCIR 601 used for broadcast stations
- MPEG-4 good for online distribution of large videos and video recorded to flash memory
- MPEG-2 used for DVDs, Super-VCDs, and many broadcast television formats
- MPEG-1 used for video CDs
- H.261
- H.263
- H.264 also known as *MPEG-4 Part 10*, or as *AVC*, used for Blu-ray Discs and some broadcast television formats
- Theora used for video on Wikipedia

Tapes

- Betacam SX, Betacam IMX, Digital Betacam, or DigiBeta — Commercial video systems by Sony, based on original Betamax technology
- D-VHS — MPEG-2 format data recorded on a tape similar to S-VHS
- D1, D2, D3, D5, D9 (also known as Digital-S) — various SMPTE commercial digital video standards
- Digital8 — DV-format data recorded on Hi8-compatible cassettes; largely a consumer format
- DV, MiniDV — used in most of today's videotape-based consumer camcorders; designed for high quality and easy editing; can also record high-definition data (HDV) in MPEG-2 format
- DVCAM, DVCPRO — used in professional broadcast operations; similar to DV but generally considered more robust; though DV-compatible, these formats have better audio handling
- DVCPRO50, DVCPROHD support higher bandwidths as compared to Panasonic's DVCPRO

- HDCAM was introduced by Sony as a high-definition alternative to DigiBeta

- MicroMV — MPEG-2-format data recorded on a very small, matchbook-sized cassette; obsolete

- ProHD — name used by JVC for its MPEG-2-based professional camcorders

Discs

- Blu-ray Disc

- DVD

- VCD

Factors of Digital Video

With digital video, four factors have to be kept in mind. These are :

- Frame rate

- Spatial Resolution

- Colour Resolution

- Image Quality

Frame Rate

The standard for displaying any type of non-film video is 30 frames per second (film is 24 frames per second). This means that the video is made up of 30 (or 24) pictures or *frames* for every second of video. Additionally these frames are split in half (odd lines and even lines), to form what are called *fields*.

Here again, there is a major difference between the way computers and television handle video. When a television set displays its analogue video signal, it displays the odd lines (the odd field) first. Then is displays the even lines (the even field). Each pair forms a frame and there are 60 of these fields displayed every second (or 30 frames per second). This is referred to as *interlaced* video.

A computer monitor, however, uses a process called *"progressive scan"* to update the screen. With this method, the screen is not broken into fields. Instead, the computer displays each line in sequence, from top to bottom. This entire frame is displayed 30 times every second. This is often called *non-interlaced* video.

Colour Resolution

This second factor is a bit more complex. Colour resolution refers to the number of colours displayed on the screen at one time. Computers deal with colour in an *RGB* (red-green-blue) format, while video uses a variety of formats. One of the most common video formats is called *YUV*. Although there is no direct correlation between RGB and YUV, they are similar in that they both have varying levels of colour depth (maximum number of colours).

Spatial Resolution

The third factor is spatial resolution - or in other words, *"How big is the picture?"*. Since PC and Macintosh computers generally have resolutions in excess of 640 by 480, most people assume that this resolution (VGA) is the video standard. It is not. As with RGB and YUV, there is no direct correlation between analogue video resolutions and computer display resolutions.

A standard analogue video signal displays a full, over scanned image without the borders common to computer screens. The National Television Standards Committee (NTSC) standard used in North America and Japanese Television uses a 768 by 484 display. The Phase Alternative system (PAL) standard for European television is slightly larger at 768 by 576. Most countries endorse one or the other, but never both.

Since the resolution between analogue video and computers is different, conversion of analogue video to digital video at times must take this into account. This can often the result in the *down-sizing* of the video and the loss of some resolution.

Image Quality

The last, and most important factor is video quality. The final objective is video that looks acceptable for your application. For some this may be 1/4 screen, 15 frames per second (fps), at 8 bits per pixel. Other require a full screen (768 by 484), full frame rate video, at 24 bits per pixel (16.7 million colours).

Compressed Video

Video compression is the process of encoding a video file in such a way that it consumes less space than the original file and is easier to transmit over the network/Internet.

It is a type of compression technique that reduces the size of video file formats by eliminating redundant and non-functional data from the original video file.

Video compression is performed through a video codec that works on one or more compression algorithms. Usually video compression is done by removing repetitive images, sounds and/or scenes from a video. For example, a video may have the same background, image or sound played several times or the data displayed/attached with video file is not that important. Video compression will remove all such data to reduce the video file size.

Once a video is compressed, its original format is changed into a different format (depending on the codec used). The video player must support that video format or be integrated with the compressing codec to play the video file.

Why Do We Need Video Compression?

One word: Bandwidth. Video is transmitted as electrical signals. These signals usually move around via air (radiowaves, microwaves, etc.) or via cables (HD-SDI, HDMI, etc.).

For simplicity's sake, it might be useful to think of cables as pipes. A pipe has a limited diameter, and there's only so much water that can flow through it, no matter how much force you apply. The limit of a broadcast system is called its bandwidth. Signals are like water, there's only so much you can push through.

In the case of wireless transmission, the 'pipe' is called the spectrum, something like an invisible bowling alley, where only a limited number of frequencies can travel (one per alley). The governments of each country limit the spectrum within their territory. The point is, even though birds have the freedom to fly any which way, your video in the air isn't allowed that freedom. Or at least, there's a cost associated with 'freedom'.

On a fundamental level, you can understand what happens when everyone wants a piece of a pipe. Ask yourself: How many people can drink from the same straw at the same time? It's the same with bandwidth.

To get more, everyone must pay more. Since this is not how humans usually deal with their money, bandwidth is distributed according to the auction methodology. Pay more, get more. Pay less, get less. Pay nothing, and be at the mercy of everyone else.

Video compression allows the efficient utilization of bandwidth by reducing file sizes. What if you could shrink a bowling ball into a marble without 'changing' its utility? Each alley becomes smaller, and you can fit more alleys in the same arena, right?

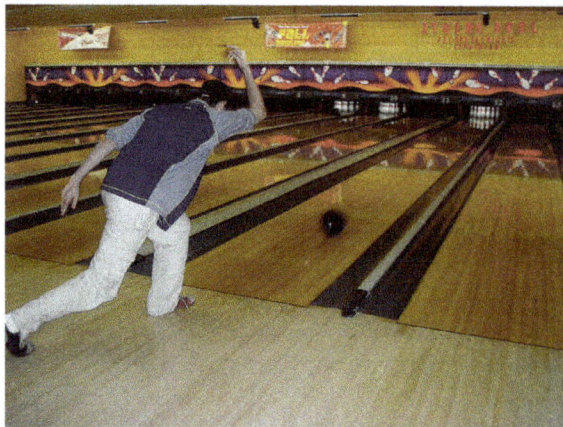

What if you could look at subsequent bowling balls and compress them into one single bowling ball? Weird? That's what interframe compression does.

Video Compression Methods

Some of the ways in which video is compressed, but are not very obvious are

- Spatial Resolution
- Color bit depth
- Chroma sub-sampling
- Gamma
- Color space
- Frame Rate, Field Rate (Temporal)
- Audio Sampling
- Audio bit depth
- Data Compression

Uncompressed Video

High-definition (HD) digital video technology has become more widely deployed in telemedicine and distance learning. Most HD videoconferencing technologies use compression to reduce the amount of data transmitted. Although it significantly lowers bandwidth requirements, compression degrades image quality and adds latency. Research and development are underway to transmit HD video without compression. These efforts can be viewed as a natural continuation of earlier ones to transmit very minimally compressed standard-definition video. If compressed HD video is relatively new and unresearched in applied telemedicine and distance learning contexts, then uncompressed video is even more so.

Three approaches to uncompressed HD videoconferencing as well as an earlier technology for delivering minimally compressed standard-definition video are described in this section. Common attributes and unique technical requirements are discussed, and the current status, benefits, and limitations of the technologies are assessed. Telemedicine and distance learning applications are identified where use of uncompressed video may be promising. The following observations are based on experiments implementing the technologies at the National Library of Medicine (NLM) and direct communication with developers. Although there are some exceptions, the technologies are too new to be widely reported in the literature.

Raw video can be sent over networks without compression. Diminished latency and improved picture quality result, but at a cost of consuming higher bandwidth. The highest image quality is attained by packetizing video from the camera directly. If the video is recorded, compression will have been applied by the camera during the recording process, and it will not look as good. Packetizing the overwhelmingly large amount of raw HD video data in real-time is accomplished with software incorporating high-efficiency algorithms and the use of more powerful computers having high-end video capture and display cards that perform much of the computation that would otherwise be done by the computer's central processing unit (CPU). The raw video resolution output from HD cameras is 1080i, but it still can be displayed on either interleaved or progressive monitors and will look better if displayed progressively. Moreover, video will be superior to current commercially available compressed HD video regardless of how it is displayed.

Echo cancellation is required for any videoconferencing system, whether video is compressed or uncompressed, to avoid audio that is transmitted from being picked up by microphones at distant sites and sent back to its source. The simplest way to provide it is to use headsets, but it is more natural to use stand-alone echo cancellation hardware, or sound cards with built-in echo cancellation and echo cancellation software. Most commercially available videoconferencing systems using compressed HD video incorporate echo cancellation, whereas the technologies for uncompressed video do not. Consequently, it may be necessary to use one of these echo cancellation strategies to avoid audio artifacts.

Recording

A standalone video recorder is a device that receives uncompressed video and stores it in either uncompressed or compressed form. These devices typically have a video output which can be used to monitor or playback recorded video. When playing back compressed video, the compressed video is uncompressed by the device before being output. Such devices may also have a communication interface, such as Ethernet or USB, which can used to exchange video files with an external computer, and in some cases control the recorder from an external computer as well.

Recording to a computer is a relatively inexpensive alternative to implementing a digital video recorder, but the computer and its video storage device (e.g., solid-state drive, RAID) must be fast enough to keep up with the high video data rate, which in some cases may be HD video or multiple video sources, or both. Due to the extreme computational and storage system performance demands of real-time video processing, other unnecessary program activity (e.g., background processes, virus scanners) and asynchronous hardware interfaces (e.g., computer networks) may be disabled, and the process priority of the recording realtime process may be increased, to avoid disruption of the recording process.

HDMI, DVI and HD-SDI inputs are available as PCI Express (partly multi-channel) or ExpressCard, USB 3.0 and Thunderbolt interface also for 2160p (4K resolution).

Software for recording uncompressed video is often supplied with suitable hardware or available for free e.g. Ingex.

Network Transmission

SMPTE 2022 is a standard for professional digital video over IP networks. The standard includes provisions for both compressed and uncompressed video formats.

Wireless interfaces such as Wireless LAN (WLAN, Wi-Fi), WiDi, and Wireless Home Digital Interface can be used to transmit uncompressed standard definition (SD) video but not HD video because the HD bit rates would exceed the network bandwidth. HD can be transmitted using higher speed interfaces such as WirelessHD and that of the Wireless Gigabit Alliance. In all cases, when video is conveyed over a network, communication disruptions or diminished bandwidth can corrupt the video or prevent its transmission.

Data Rates

Uncompressed video has a constant bitrate that is based on pixel representation, image resolution, and frame rate:

> data rate = color depth * vertical resolution * horizontal resolution * refresh frequency

For example:

- 24-bit, 1080i @ 60 fps: 24 × 1920×1080 × 60/2 = 1.49 Gbit/s

- 24-bit, 1080p @ 60 fps: 24 × 1920×1080 × 60 = 2.98 Gbit/s.

NTSC

```
8-bit, 720×480 @ 29.97 fps = 20 MB/s, or 70 GB/h

10-bit, 720×480 @ 29.97 fps = 27 MB/s, or 94 GB/h
```

PAL

```
8-bit, 720×576 @ 25 fps = 20 MB/s, or 70 GB/h

10-bit, 720×576 @ 25 fps = 26 MB/s, or 93 GB/h
```

720p

```
8-bit, 1280×720 @ 59.94 fps = 105 MB/s, or 370 GB/h

10-bit, 1280×720 @ 59.94 fps = 140 MB/s, or 494 GB/h
```

1080i and 1080p

> *8-bit, 1920×1080 @ 24 fps = 95 MB/s, or 334 GB/h*
>
> *10-bit, 1920×1080 @ 24 fps = 127 MB/s, or 445 GB/h*
>
> *8-bit, 1920×1080 @ 25 fps = 99 MB/s, or 348 GB/h*
>
> *10-bit, 1920×1080 @ 25 fps = 132 MB/s, or 463 GB/h*
>
> *8-bit, 1920×1080 @ 29.97 fps = 119 MB/s, or 417 GB/h*
>
> *10-bit, 1920×1080 @ 29.97 fps = 158 MB/s, or 556 GB/h*

1080i and 1080p RGB (4:4:4)

> *10-bit, 1280×720p @ 60 fps = 211 MB/s, or 742 GB/h*
>
> *10-bit, 1920×1080 @ 24 fps = 190 MB/s, or 667 GB/h*
>
> *10-bit, 1920×1080 @ 50i = 198 MB/s, or 695 GB/h*
>
> *10-bit, 1920×1080 @ 60i = 237 MB/s, or 834 GB/h*

4k (3840x2160)

> *8-bit, 3840x2160 @ 24 fps = 380 MB/s, or 1.33 TB/h*
>
> *8-bit, 3840x2160 @ 30 fps = 475 MB/s, or 1.67 TB/h*
>
> *8-bit, 3840x2160 @ 60 fps = 950 MB/s, or 3.33 TB/h*

8k (7680x4320)

> *8-bit, 7680x4320 @ 24 fps = 1.52 GB/s, or 5.33 TB/h*
>
> *8-bit, 7680x4320 @ 30 fps = 1.9 GB/s, or 6.67 TB/h*
>
> *8-bit, 7680x4320 @ 60 fps = 3.8 GB/s, or 13.33 TB/h*

Advantages of Uncompressed Video

Why is uncompressed video the ideal for the professional videographer? In the simplest terms, uncompressed video recording produces higher-quality images. Compressed video often has issues with slightly off-color gradients and electronically generated backgrounds, which are created during the compression process. Compression, while not always noticeable in small doses, can become a major distraction when overused. Typically, compression is utilized for representing redundancies as a single unit.

Examples of Codec Packs

Nowadays there is a good number of media players like VLC that can handle an astonishing number of video and audio file with various codecs. However, some codecs might need additional software for their playback.

K-Lite Codec Pack

K-Lite Codec Pack is a well-known free codec compilation software. It's easy to install and use. The tool contains an impressive codec library which gets regular updates. You can choose among four versions of this product.

- K-lite Codec Pack Basic contains only the main codecs you might need.

- K-lite Codec Pack Standard provides codecs to all most popular audio and video files.

- K-Lite Codec Pack Full gives more specialized codec libraries.

- K-Lite Codec Pack Mega is for those who'd like to have everything that is possible.

Recently K-Lite Codec Pack has added full decoding support for the new HEVC (H.265) video codec.

X Codec Pack

X Codec Pack (formerly XP Codec Pack) is a great alternative to K-Lite in case you don't want to use it for any reason. Like the previous software, X Codec Pack contains all codecs that you might need to play popular and rare videos files. There are two drawbacks here. First of all, the tool is that it doesn't get regular updates. Thus, you might need to wait for some time for new codecs to be added. The 2nd issue is that a full codec pack may interfere with your media player and cause problems in its performance.

References

- Raake, Alexander; Egger, Sebastian (2014). Quality of Experience. T-Labs Series in Telecommunication Services. Springer, Cham. pp. 11–33. doi:10.1007/978-3-319-02681-7_2. ISBN 9783319026800

- Winkler, Stefan. "The evolution of video quality measurement: from PSNR to hybrid metrics". IEEE Transactions on Broadcasting. doi:10.1109/TBC.2008.2000733

- Shahid, Muhammad; Rossholm, Andreas; Lövström, Benny; Zepernick, Hans-Jürgen (2014-08-14). "No-reference image and video quality assessment: a classification and review of recent approaches". EURASIP Journal on Image and Video Processing. 2014: 40. doi:10.1186/1687-5281-2014-40. ISSN 1687-5281

- Raake, Alexander, (2006). Speech quality of VoIP : assessment and prediction. Wiley InterScience (Online service). Chichester, England: Wiley. ISBN 9780470030608. OCLC 85785040

- Saad, M. A.; Bovik, A. C.; Charrier, C. (March 2014). "Blind Prediction of Natural Video Quality". IEEE Transactions on Image Processing. 23(3): 1352–1365. doi:10.1109/tip.2014.2299154. ISSN 1057-7149

- "SERIES H: AUDIOVISUAL AND MULTIMEDIA SYSTEMS : Infrastructure of audiovisual services – Coding of moving video : Advanced video coding for generic audiovisual services". Itu. int. Retrieved 6 January 2015

- Möller, Sebastian, (2000). Assessment and Prediction of Speech Quality in Telecommunications. Boston, MA: Springer US. ISBN 9781475731170. OCLC 851800613

- Saad, M. A.; Bovik, A. C.; Charrier, C. (March 2014). "Blind Prediction of Natural Video Quality". IEEE Transactions on Image Processing. 23(3): 1352–1365. doi:10.1109/tip.2014.2299154. ISSN 1057-7149

- Rowan Trollope (2013-10-30). "Open-Sourced H.264 Removes Barriers to WebRTC". Cisco. Retrieved 2016-05-23

4

An Introduction to Video Processing

Video processing is a form of signal processing, which implements video filters for getting video streams or video files, for use in TV sets, DVDs, VCRs, video players, etc. This chapter has been carefully written to provide an easy understanding of the varied aspects of video processing such as scan conversion, video capture, deinterlacing, color grading, motion interpolation, etc.

Video Processing

Video processing uses hardware, software, and combinations of the two for editing the images and sound recorded in video files. Extensive algorithms in the processing software and the peripheral equipment allow the user to perform editing functions using various filters. The desired effects can be produced by editing frame by frame or in larger batches.

Most modern personal computers come with software that allows the user to compile images and videos, edit images, and create videos on a limited level. Storyboards allow the addition of audio files and the adjustment of visual images, transitions, and audio files, which, together, determine the overall length of the video. Videographers, electrical engineers, and computer science professionals use programs that are capable of a wider range of functions. Signal processing usually involves applying a combination of prefilters, intrafilters, and post filters.

Video files are obtained from the recording device using a universal standard bus (USB) cable or firewire attachment. The files are then loaded into a computer software program or peripheral device. Before applying the filters used in video processing, certain programs require information for the optimization framework. This information allows the program to calculate the horizontal and vertical image gradients, determine the desired filter gradients, and establish function parameters.

Prefilters used in video processing might include contrast changes, deflicking, and noise elimination along with pixel size conversions. Contrast changes allow the processor to highlight particular areas of an image, change the lighting perspective, and darken or lighten images. Deflicking eliminates camera motion or uneven lighting effects that produce flickering on the video. Noise elimination removes artifacts, including lines or other textured effects that reduce image clarity. Using size conversions, users might change a video from 720 pixels to 1,080 pixels, crop the size of the video, or reposition the video on a background.

Processing videos using intrafilters allows users to deblock, or apply techniques that change the image quality. Deblocking removes blocking artifacts, sometimes acquired by compressed files, that reduce image clarity. Using the calculated gradient aspects of images, filters might sharpen out of focus images, apply highlighting around specified areas of an image, or add graphics and text to a video. Filters can also change entire color schemes or vary the colors within an image.

Deinterlacing is a post filter that is frequently used in video processing. When video recorders capture images, the images can overlap or interlace over each other. This creates artifacts that might include blurred images, a checkerboard effect, or lines that become visible during playback. Deinterlacing programs eliminate these problems by combining frames and allowing progressive scanning without these visual disturbances.

Video Processor

The electronic definition of a processor as it relates to this discussion is a device that transforms a video signal or signals into something acceptable for another use.

There are many other names that describe a Video Processor (referred to hereafter as VP) such as Display Wall Processor, Display Wall Controller, Wall Controller, Wall Processor, Video Server, Data Wall Processor, Data Wall Server, Multiviewer, etc.

Types of Video Processor

We will refer to all of these as VP in this white paper. There are 3 basic categories of Video Processors:

1) Single input, multiple outputs.

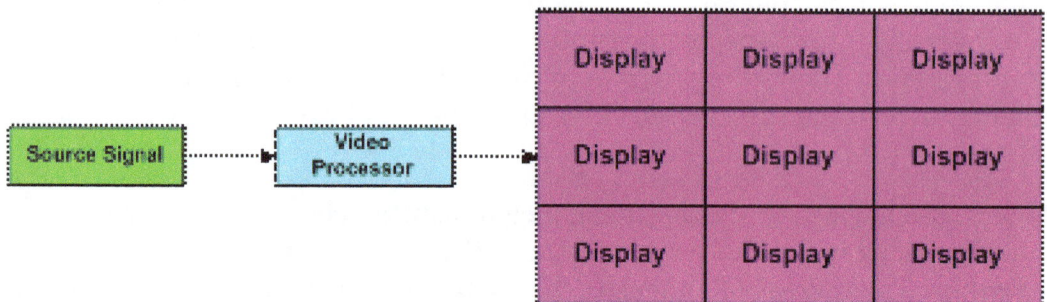

2) Multiple inputs, single output

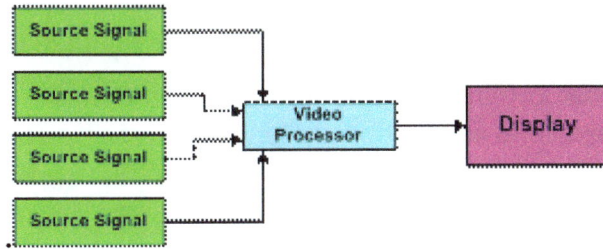

3) Multiple inputs, multiple outputs.

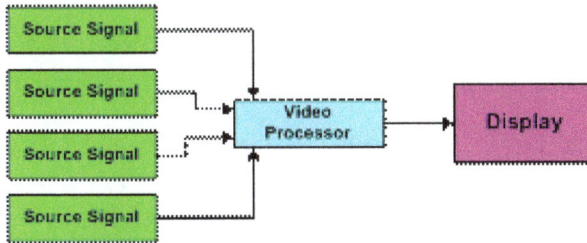

Most Video Processors are hardware devices that are built to order because the number of inputs and outputs required are specific to a given installation's needs.

Uses

Video processors are often combined with video scalers to create a video processor that improves the apparent definition of video signals. They perform the following tasks:

- deinterlacing
- aspect ratio control
- digital zoom and pan
- brightness/contrast/hue/saturation/sharpness/gamma adjustments
- frame rate conversion and inverse-telecine
- color point conversion (601 to 709 or 709 to 601)
- color space conversion (YP_BP_R/YC_BC_R to RGB or RGB to YP_BP_R/YC_BC_R)
- mosquito noise reduction
- block noise reduction
- detail enhancement
- edge enhancement
- motion compensation
- primary and secondary color calibration (including hue/saturation/luminance controls independently for each)

These can either be in chip form, or as a stand-alone unit to be placed between a source device (like a DVD player or set-top-box) and a display with less-capable processing. The most widely recognized video processor companies in the market are:

- Genesis Microchip (with the FLI chipset – was Genesis Microchip, STMicroelectronics completes acquisition of Genesis Microchip on January 25, 2008)

- Sigma Designs (with the VXP chipset – was Gennum, Sigma Designs purchased the Image Processing group from Gennum on February 8, 2008)

- Integrated Device Technology (with the HQV chipset and Teranex system products – was Silicon Optix, IDT purchased SO on October 21, 2008)

- Silicon Image (with the VRS chipset and DVDO system products - was Anchor Bay Technologies, Silicon Image purchased ABT on February 10, 2011)

All of these companies' chips are in devices ranging from DVD upconverting players (for Standard Definition) to HD DVD/Blu-ray Disc players and set-top boxes, to displays like plasmas, DLP (both front and rear projection), LCD (both flat-panels and projectors), and LCOS/"SXRD". Their chips are also becoming more available in stand alone devices.

Deflicking

Brightness flicker is an annoying effect. It can be seen in video digitized from old film or captured by low-quality camera equipment, and it is apparent under some artificial lighting conditions when the frame rate has not been adjusted with respect to the flicker frequency of the lighting.

In video processing, deflicking is a filtering operation applied to brightness flicker in video to improve visual quality. The filter aims to improve the appearance of movies.

The main idea is to smooth image brightness between series of the same scene frames.

The deflicking filter is usually used in video camera (for normalizing picture), used for postprocessing of captured video, and for restoration of video from old films.

Scan Conversion

A fundamental operation that is used extensively in computer graphics and visualization is the process of scan conversion or rasterization. Given a polygon in image space, this process determines the pixels that intersect the polygon. This process is utilized in visible-surface algorithms, incremental-shading techniques, polygon-fill algorithms, ray-tracing-acceleration algorithms, and a number of other tasks that are critical to the understanding of the computer graphics field.

- 2D or 3D objects in real world space are made up of graphic primitives such as points, lines, circles and filled polygons.

- These picture components are often defined in a contiguous space at a higher level of abstraction than individual pixels in the discrete image space.

- For instance, a line is defined by its two endpoints and the line equation while a circle is defined by its radius, centre position, and the circle equation.

- It is the responsibility of the graphics system or the application program to convert each primitive from its geometric definition into a set of pixels that makes up the primitive in the image space.

- This conversion task is generally referred to as scan-conversion or rasterization.

The concept of SCAN CONVERSION is important in raster graphics since any shape which is to be drawn into the frame buffer must first be decomposed into pixels lying in a regular raster grid pattern. Typical shapes which the graphics display generator may scan convert are POINTS, LINES, RECTANGLES, CIRCLES, CONICS, DISKS, CHARACTERS, SPECIAL SYMBOLS, ICONS, BITMAPS, PATTERNS, and POLYGONS, as well as REGIONS defined by or bounded by pixel values in the frame buffer itself. Some of the perceived power of a raster graphics display depends on the primitive vocabulary of scan converted shapes: the more that are handled by the display processor, the better.

Scan Conversion of Trapezoids in Device Space

Scan Conversion is the process of finding the screen pixels that intersect a polygon. To do this, we find it convienent to move to a copy of image space that is scaled to closely correspond to the pixels in our display window. This space, commonly called device space is parameterized so that the lower-left-hand corner is at $(0,0)$, and so that the pixels can all be indexed by integers.

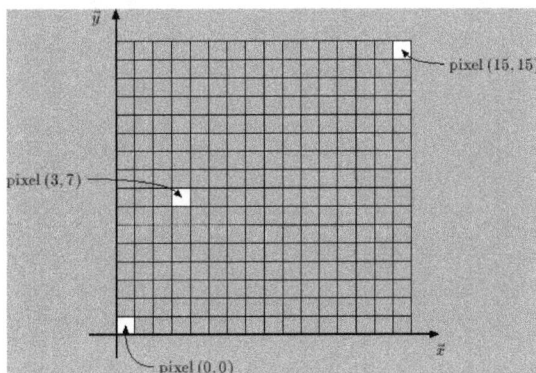

To scan convert a polygon in this space, we will split the polygon into a set of trapezoids and then scan convert each trapezoid. Each trapezoid will be of a special form where the top and bottom edges of the trapezoid are parallel to the scanlines (i.e., of a constant y value). We will also consider "degenerate" trapezoids - triangles - which have either the top of bottom edge of zero length. In the following illustration, the polygon is split into five trapezoids. The top and bottom trapezoids are actually triangles.

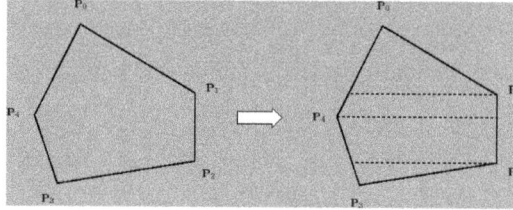

The union of all pixels that intersect the set of trapezoids will be the set of pixels that intersect the polygon.

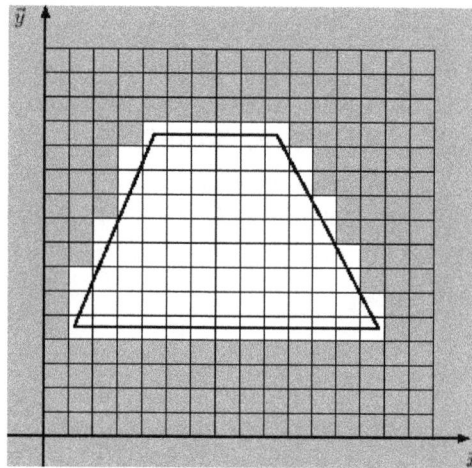

The idea here is easy. We will establish an "edge tracker" which follows the endpoints of the lines formed by intersecting each scanline with the trapezoid.

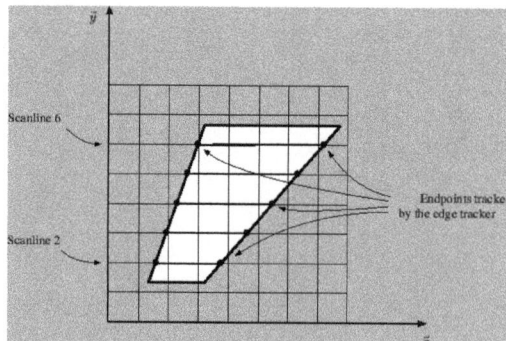

This edge tracker can be easily defined as an simple data structure which is initialized for the scanline at the top of each trapezoid and updated for each subsequent scanline.

Mechanisms and Methods

Scan conversion involves changing the picture information data rate and wrapping the new picture in appropriate synchronization signals. There are two distinct methods for changing a picture's data rate:

- Analog Methods (Non Retentive, Memory-less or Real Time Method)

This conversion is done using large numbers of delay cells and is appropriate for analog video. It may also be performed using a specialized scan converter vacuum tube. In this case polar coordinates (angle and distance) data from a source such as a radar receiver, so that it can be displayed on a raster scan (TV type) display.

- Digital methods (Retentive or buffered method)

In this method, a picture is stored in a line or frame buffer with n1 speed (data rate) and is read with n2 speed, several picture processing techniques are applicable when the picture is stored in buffer memory including kinds of interpolation from simple to smart high order comparisons, motion detection and ... to improve the picture quality and prevent the conversion artifacts.

Ways to Realize

The process in practice is applicable only using integrated circuits in LSI and VLSI scales. Timing, interference between digital and analog signals, clocks, noise and exact synchronization have important roles in the circuit. Digital conversion method needs the analog video signal to be converted to digital data at the first step. A scan converter can be made in its basic structure using some high speed integrated circuits as a circuit board however there are some integrated circuits which perform this function plus other picture processing functions like scissoring, change of aspect ratio and an easy to use example was SDA9401.

Some Examples

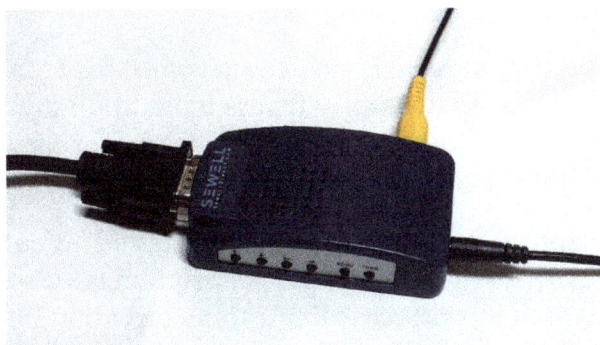

A VGA to TV scan converter box like this turns enhanced-definition or high-definition signals into standard-definition signals.

Up conversion (interpolation):

- In many LCD monitors there is a native picture mode, however the monitor can display different graphical modes using a scan converter.

- In a 100 Hz/120 Hz analog TV, there is a scan converter circuit which converts the vertical frequency (refresh rate) from standard 50/60 Hz to 100/120 Hz to achieve a low level of flicker which is important in large screen (high inch) TVs.

- An external TV card receives the TV signals and converts them to VGA or SVGA format to display on monitor.

Down conversion (decimation):

- Many graphic cards have output for standard-definition television. Here there is a conversion from computer graphical modes to TV standard formats.

- Other graphic cards lack an SDTV output, but their VGA outputs can still be connected to an SDTV through an external scan converter (pictured).

Scan conversion serves as a bridge between TV and computer graphics technology.

Scan Conversion is a procedure that is used repeatedly in computer graphics algorithms. It is a simple procedure, designed around an edge-tracking paradigm, which can be implemented by adding simply-calculated increments onto base values. One of the primary uses of the algorithm is to establish depth (z) values at each pixel for polygons in the scene, this enables us to retain only the visible polygons in our final renderings. However, we can also utilize quantities relating to color, texture, and parameterization as information in our endpoint nodes and increment them as well.

Video Capture

A video capture is a digitized version of an external video feed. Capturing video usually requires encoding or post-production software in addition to whatever hardware is being used to transmit the original feed into its digital file format (which can include a tape deck, digital storage or a video camera).

Broadly speaking, a capture is essentially a quantized and/or compressed version of some external source. Within the scope of that description, a video capture can include a camera recording as well as that recording's transformation into an encoded, playable file. Typically, however, in the realm of video production and post production, the capture process describes when an external video feed (such as an analog signal) is digitized.

Within the context of digital video production and encoding, video captures can involve tape-to-file capturing as well as capturing from a variety of other media sources (such as the camera itself). As video technology becomes more advanced and consolidated, however, all phases of the video production pipeline (from original footage to deliverable media) are converging into turnkey, often mobile devices. Many modern consumer smartphones, for example, are capable of shooting, editing and encoding video all within their own operating system.

Video Capture Device

A video capture device is a piece of hardware that lets you transfer audio and video from a VCR, camcorder, or other device, to your computer so that it can be stored on a hard drive, whether for editing or just general archival purposes.

For example, a video capture device can be used to convert VHS tapes to a digital video format that you can then put on a DVD, upload to YouTube, edit on your computer, etc.

While many people place TV tuners and video capture devices in the same category, they are not equal. There are many overlaps in terms of what they do, but video capture devices will not tune TV channels nor can most be used as a tuner without external equipment.

Special electronic circuitry is required to capture video from analog video sources. At the system level this function is typically performed by a dedicated video capture device. Such devices typically employ integrated circuit video decoders to convert incoming video signals to a standard digital video format, and additional circuitry to convey the resulting digital video to local storage or to circuitry outside the video capture device, or both. Depending on the device, the resulting video stream may be conveyed to external circuitry via a computer bus (e.g., PCI/104 or PCIe) or a communication interface such as USB, Ethernet or WiFi, or stored in mass-storage memory in the device itself (e.g., digital video recorder).

| A PCIe 2-port video capture card (Datapath Vision-RGB-E2s) | A Mini PCIe card that simultaneously captures 8 video and 8 audio signals (Sensoray 1012) | A low-cost, consumer-grade USB audio/video capture device (Reddo Videosieppari) |

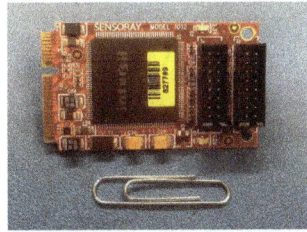

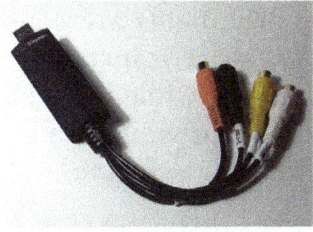

Video Capture Device for Live Streaming

Most modern laptops have a webcam which makes live broadcast available to a software on that system. Within the camera, there is a chip that converts the analog visual world into digital data. The system sees the webcam as a video device which can be used by software such as Skype or Google Hangouts. It can also be used by most live broadcast software, like Telestream Wirecast or vMix.

The unfortunate fact is that your little webcam doesn't give you the ability to connect an external video source. This is where a third-party capture device comes to the rescue. It allows you to connect an external video feed to your computer, and makes that source available as a audio/video capture device selection in your software.

Deinterlacing

To prevent flicker and reducing transmission bandwidth, all analog camcorders, VCRs, broadcast television systems use interlaced scan. So, deinterlacing video is in practice the process of converting interlaced video into progressive video.

Need to Deinterlace Video

Most modern displays, such as LCD, DLP and plasma displays, only work in deinterlace mode, because they are fixed-resolution displays and only support progressive scan. In order to display interlaced signal on such displays, the two interlaced fields must be combined into one progressive frame with a deinterlacing process.

Background

Both video and photographic film capture a series of frames (still images) in rapid succession; however, television systems read the captured image by serially scanning the image sensor by lines (rows). In analog television, each frame is divided into two consecutive fields, one containing all even lines, another with the odd lines. The fields are captured in succession at a rate twice that of the nominal frame rate. For instance, PAL and SECAM systems have a rate of 25 frames/s or 50 fields/s, while the NTSC system delivers 29.97 frames/s or 59.94 fields/s. This process of dividing frames into half-resolution fields at double the frame rate is known as *interlacing*.

Since the interlaced signal contains the two fields of a video frame shot at two different times, it enhances motion perception to the viewer and reduces flicker by taking advantage of the persistence of vision effect. This results in an effective doubling of time resolution as compared with non-interlaced footage (for frame rates equal to field rates). However, interlaced signal requires a display that is natively capable to show the individual fields in a sequential order, and only traditional CRT-based TV sets are capable of displaying interlaced signal, due to the electronic scanning and lack of apparent fixed resolution.

Most modern displays, such as LCD, DLP and plasma displays, are not able to work in interlaced mode, because they are fixed-resolution displays and only support progressive scanning. In order to display interlaced signal on such displays, the two interlaced fields must be converted to one progressive frame with a process known as *de-interlacing*. However, when the two fields taken at different points in time are re-combined to a full frame displayed at once, visual defects called *interlace artifacts* or *combing* occur with moving objects in the image. A good deinterlacing algorithm should try to avoid interlacing artifacts as much as possible and not sacrifice image quality in the process, which is hard to achieve consistently. There are several techniques available that extrapolate the missing picture information, however they rather fall into the category of intelligent frame creation and require complex algorithms and substantial processing power.

Deinterlacing techniques require complex processing and thus can introduce a delay into the video feed. While not generally noticeable, this can result in the display of older video games lagging behind controller input. Many TVs thus have a "game mode" in which minimal processing is done in order to maximize speed at the expense of image quality. Deinterlacing is only partly responsible for such lag; scaling also involves complex algorithms that take milliseconds to run.

Progressive Source Material

Interlaced video can carry progressive scan signal, and deinterlacing process should consider this as well.

Typical movie material is shot on 24 frames/s film; when converting film to interlaced video using telecine, each film frame can be presented by two progressive segmented frames (PsF). This format does not require complex deinterlacing algorithm because each field contains a part of the very same progressive frame. However to match 50 field interlaced PAL/SECAM or 59.94/60 field interlaced NTSC signal, frame rate conversion should be performed using various "pulldown" techniques; most advanced TV sets can restore the original 24 frame/s signal using an inverse telecine process. Another option is to speed up 24-frame film by 4% (to 25 frames/s) for PAL/SECAM conversion; this method is still vastly used for DVDs, as well as television broadcasts (SD & HD) in the PAL markets.

DVDs can either encode movies using one of these methods, or store original 24 frame/s progressive video and use MPEG-2 decoder tags to instruct the video player on how to convert them to the interlaced format. Most movies on Blu-ray discs have preserved the original non interlaced 24 frame/s motion film rate and allow output in the progressive 1080p24 format directly to display devices, with no conversion necessary.

Some 1080i HDV camcorders also offer PsF mode with cinema-like frame rates of 24 or 25 frame/s. The TV production can also use special film cameras which operate at 25 or

30 frame/s; such material does not need framerate conversion for broadcasting in the intended video system format.

Deinterlacing Methods

Deinterlacing requires the display to buffer one or more fields and recombine them into full frames. In theory this would be as simple as capturing one field and combining it with the next field to be received, producing a single frame. However, the originally recorded signal was produced as a series of fields, and any motion of the subjects during the short period between the fields is encoded into the display. When combined into a single frame, the slight differences between the two fields due to this motion results in a "combing" effect where alternate lines are slightly displaced from each other.

There are various methods to deinterlace video, each producing different problems or artifacts of its own. Some methods are much cleaner in artifacts than other methods.

Most deinterlacing techniques can be broken up into three different groups all using their own exact techniques. The first group are called *field combination deinterlacers*, because they take the even and odd fields and combine them into one frame which is then displayed. The second group are called *field extension deinterlacers*, because each field (with only half the lines) is extended to the entire screen to make a frame. The third type uses a combination of both and falls under the banner of *motion compensation* and a number of other names.

Modern deinterlacing systems therefore buffer several fields and use techniques like edge detection in an attempt to find the motion between the fields. This is then used to interpolate the missing lines from the original field, reducing the combing effect.

Field Combination Deinterlacing

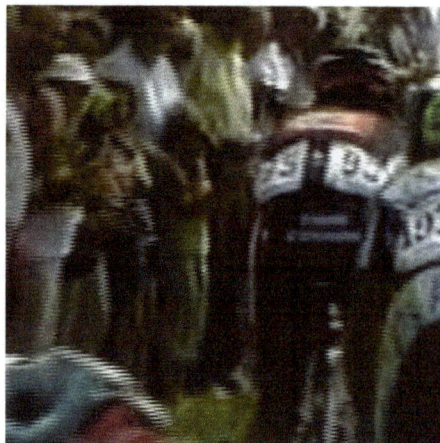

Weaving

- Weaving is done by adding consecutive fields together. This is fine when the image hasn't changed between fields, but any change will result in artifacts known as "combing," when the pixels in one frame do not line up with the pixels in the other, forming a jagged edge. This technique retains the full vertical resolution at the expense of half the temporal resolution (motion).

Blending

- Blending is done by *blending*, or *averaging* consecutive fields to be displayed as one frame. Combing is avoided because the images are on top of each other. This instead leaves an artifact known as ghosting. The image loses vertical resolution and temporal resolution. This is often combined with a vertical resize so that the output has no numerical loss in vertical resolution. The problem with this is that there is a quality loss, because the image has been downsized then upsized. This loss in detail makes the image look softer. Blending also loses half the temporal resolution since two motion fields are combined into one frame.

- Selective blending, or *smart blending* or *motion adaptive blending*, is a combination of weaving and blending. As areas that haven't changed from frame to frame don't need any processing, the frames are woven and only the areas that need it are blended. This retains the full vertical resolution and half the temporal resolution, and it has fewer artifacts than weaving or blending because of the selective combination of both techniques.

- Inverse Telecine: Telecine is used to convert a motion picture source at 24 frames per second to interlaced TV video in countries that use NTSC video system at 30 frames per second. Countries which use PAL at 25 frames per second do not use Telecine since motion picture sources are sped up 4% to achieve the needed 25 frames per second. If Telecine was used then it is possible to reverse the algorithm to obtain the original non-interlaced footage, which has a slower frame rate. In order for this to work, the exact telecine pattern must be known or guessed. Unlike most other deinterlacing methods, when it works, inverse telecine can perfectly recover the original progressive video stream.

- Telecide-style algorithms: If the interlaced footage was generated from progressive frames at a slower frame rate (e.g. "cartoon pulldown"), then the exact original frames can be recovered by copying the missing field from a matching previous/next frame. In cases where there is no match (e.g. brief cartoon sequences with an elevated frame rate), then the filter falls back on another deinterlacing method such as blending or line-doubling. This means that the worst case for Telecide is occasional frames with ghosting or reduced resolution. By contrast, when more sophisticated motion-detection algorithms fail, they can introduce pixel artifacts that are unfaithful to the original material. For telecine video, decimation can be applied as a post-process to reduce the frame rate, and this combination is generally more robust than a simple inverse telecine, which fails when differently interlaced footage is spliced together.

Field Extension Deinterlacing

Half-sizing

Half-sizing displays each interlaced field on its own, resulting in a video with half the vertical resolution of the original, unscaled. While this method retains all vertical resolution and all temporal resolution it is understandably not used for regular viewing because of its false aspect ratio. However, it can be successfully used to apply video filters which expect a noninterlaced frame, such as those exploiting information from neighbouring pixels (e.g., sharpening).

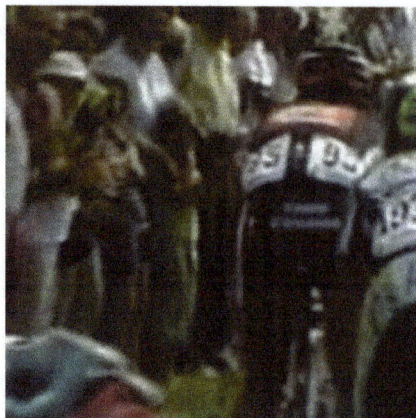

Line doubling

Line doubling takes the lines of each interlaced field (consisting of only even or odd lines) and doubles them, filling the entire frame. This results in the video having a frame rate identical to the field rate, but each frame having half the vertical resolution, or resolution equal to that of each field that the frame was made from. Line doubling prevents combing artifacts but causes a noticeable reduction in picture quality since each frame displayed is doubled and really only at the original half field resolution. This is noticeable mostly on stationary objects since they appear to bob up and down. These techniques are also called *bob deinterlacing* and *linear deinterlacing* for this reason. Line doubling retains horizontal and temporal resolution at the expense of vertical resolution and bobbing artifacts on stationary and slower moving objects. A variant of this method discards one field out of each frame, halving temporal resolution.

Line doubling is sometimes confused with deinterlacing in general, or with interpolation (image scaling) which uses spatial filtering to generate extra lines and hence reduce the visibility of pixelation on any type of display. The terminology 'line doubler' is used more frequently in high end consumer electronics, while 'deinterlacing' is used more frequently in the computer and digital video arena.

Motion Detection

Best picture quality can be ensured by combining traditional field combination methods (weaving and blending) and frame extension methods (bob or line doubling) to create a high quality progressive video sequence; the best algorithms would also try to predict the direction and the amount of image motion between subsequent sub-fields in order to better blend the two subfields together.

One of the basic hints to the direction and amount of motion would be the direction and length of combing artifacts in the interlaced signal. More advanced implementations would employ algorithms similar to block motion compensation used in video compression; deinterlacers that use this technique are often superior because they can use information from many fields, as opposed to just one or two. This requires powerful hardware to achieve realtime operation.

For example, if two fields had a person's face moving to the left, weaving would create combing, and blending would create ghosting. Advanced motion compensation (ideally) would see that the face in several fields is the same image, just moved to a different position, and would try to detect direction and amount of such motion. The algorithm would then try to reconstruct the full detail of the face in both output frames by combining the images together, moving parts of each subfield along the detected direction by the detected amount of movement.

Motion compensation needs to be combined with scene change detection, otherwise it will attempt to find motion between two completely different scenes. A poorly implemented motion compensation algorithm would interfere with natural motion and could lead to visual artifacts which manifest as "jumping" parts in what should be a stationary or a smoothly moving image.

Place where Deinterlacing is Performed

Deinterlacing of an interlaced video signal can be done at various points in the TV production chain.

Progressive Media

Deinterlacing is required for interlaced archive programs when the broadcast format or media format is progressive, as in EDTV 576p or HDTV 720p50 broadcasting, or mobile DVB-H broadcasting; there are two ways to achieve this.

- *Production* – The interlaced video material is converted to progressive scan during program production. This should typically yield the best possible quality, since videographers have access to expensive and powerful deinterlacing equipment and software and can deinterlace at the best possible quality, probably manually choosing the optimal deinterlacing method for each frame.

- *Broadcasting* – Real-time deinterlacing hardware converts interlaced programs to progressive scan immediately prior to broadcasting. Since the processing time is constrained by the frame rate and no human input is available, the quality of conversion is most likely inferior to the pre-production method; however, expensive and high-performance deinterlacing equipment may still yield good results when properly tuned.

Interlaced Media

When the broadcast format or media format is interlaced, real-time deinterlacing should be performed by embedded circuitry in a set-top box, television, external video processor, DVD or DVR player, or TV tuner card. Since consumer electronics equipment is typically far cheaper, has considerably less processing power and uses simpler algorithms compared to professional deinterlacing equipment, the quality of deinterlacing may vary broadly and typical results are often poor even on high-end equipment.

Using a computer for playback and/or processing potentially allows a broader choice of video players and/or editing software not limited to the quality offered by the embedded consumer electronics device, so at least theoretically higher deinterlacing quality is possible – especially if the user can pre-convert interlaced video to progressive scan before playback and advanced and time-consuming deinterlacing algorithms (i.e. employing the "production" method).

However, the quality of both free and commercial consumer-grade software may not be up to the level of professional software and equipment. Also, most users are not trained in video production; this often causes poor results as many people do not know much about deinterlacing and are unaware that the frame rate is half the field rate. Many codecs/players do not even deinterlace by themselves and rely on the graphics card and video acceleration API to do proper deinterlacing.

Concerns Over Effectiveness

The European Broadcasting Union has argued against the use of interlaced video in production and broadcasting, recommending 720p 50 fps (frames per second) as current production format and working with the industry to introduce 1080p50 as a future-proof production standard which offers higher vertical resolution, better quality at lower bitrates, and easier conversion to other formats such as 720p50 and 1080i50. The main argument is that no matter how complex the deinterlacing algorithm may be, the artifacts in the interlaced signal cannot be completely eliminated because some information is lost between frames.

Yves Faroudja, the founder of Faroudja Labs and Emmy Award winner for his achievements in deinterlacing technology, has stated that "interlace to progressive does not work" and advised against using interlaced signal.

To Deinterlace or Not

Nowadays, CRT TVs are no longer being manufactured. So it's safe to assume that whatever content you wish to broadcast to your audience, the (overwhelming) majority of it will be consumed over flat panel screens (LCDs, tablets, PCs, Plasmas, etc.).

Given that flat panel TVs cannot play interlaced content, if you have interlaced content (e.g. 576i, 480i, 1080i) – it will be deinterlaced by those devices in real-time.

Now if you only have Hollywood movies, i.e. content that was shot with film cameras, it will be progressive and not interlaced, so you have no problem.

But if you are getting a lot of video content, which was shot on interlaced video cameras, either SD or HD (1080i), you have to deal with this annoying interlacement issue.

Everyone in the industry pretty much agrees by now that interlacing should be eradicated for good, as it is an old analog technology created because of ancient equipment limitations.

The majority of video cameras today are supporting purely progressive shooting, so 1080p becomes quite common.

If you choose to deinterlace, you also have to consider the quality of deinterlacing. When it comes to software, there are quick deinterlacers which give reasonable results, and there are slow, motion-compensated algorithms which produce superior quality (although it is questionable what percentage of your audience can actually tell the difference).

Another point to consider is how efficiently your encoder can handle interlaced streams. If your encoder is more efficient in progressive encoding, then run it through a reasonable deinterlacer, one that doesn't add too much to the encoding time. It would give better results than encoding inefficiently in interlaced mode.

Keep in mind that when deinterlacing, as with ANY video processing, the image is being processed and will be forever changed from its source. If you can afford the storage, keep your masters untouched for future repurposing.

Telecine

Telecine is the conversion process of motion picture film into video. Telecine helps in the viewing of motion pictures with standard video devices such as televisions or computers. Telecine initially only dealt with film-to-video conversion, but with the advent of digital televisions, telecine algorithms were incorporated into devices such as televisions and DVDs to include frame rate conversion, upconversion and deinterlacing. Telecine has the ability to reframe the shots, thus the film can undergo dramatic changes. However with the advent of scanners, telecine is less popular than it was previously.

Telecine is also known as cinema pulldown 3:2 or 3:2 pulldown.

Telecine is performed in a color suite. The most challenging part in telecine is in the synchronization of the film motion with the electronic video signal. It is easy to perform when the film is at the same frame rate, but if that is not the case, a complex process is required for changing the frame rate to establish the synchronization. The basic idea behind telecine is to capture each and every frame and store them. For this, the movie is to be recorded at a frame rate of 16/18 fps. However the normal video speed is 25-30 fps and is thus adjusted with the help of a procedure called the pull-down technique. Telecine can also be reversed using a process called inverse-telecine.

Telecine, Pull-Down, and Reverse Telecine

The following sections describe methods for embedding and extracting 24p video in different formats. Some of these techniques are based on existing film-to-video methods, and some are newer approaches. The basic technique for transferring film to video, uses a process called pull-down to map 23.98 fps film to 29.97 fps interlaced video. Once the video is captured on disk, software can perform reverse telecine, or reverse pull-down, to restore the original 23.98 fps film frame rate.

In progressive digital video systems such as 720p60 DVCPRO HD video, a similar process can be performed in-camera to map 23.98 fps to 59.94 fps, but entire frames are duplicated instead of fields. During or after capture, the duplicate frames are removed. A camcorder or deck that performs duplicate frame insertion can add metadata (known as flags) that inform software when to remove or ignore duplicate frames.

Standard 3:2 Pull-Down

Also known as *2:3:2:3 pull-down*, this is the standard telecine method of transferring film to NTSC video. The film is slowed by 0.1 percent (a factor of 1000/1001) from 24 fps to 23.98 fps, and then each film frame is transferred to interlaced video in a repeating 2:3:2:3 field pattern.

In the illustration, film frames A, B, and D are mapped to video frames 1, 2, and 5. However, because film frame C is split into two fields across video frames 3 and 4, pull-down removal requires deinterlacing, which is more processor-intensive than removal of pull-down patterns such as advanced (2:3:3:2) pull-down.

Pull-down removal typically requires manual identification of the A frame in the pattern, which you can identify visually by moving frame by frame through your footage until you recognize the pull-down pattern.

If you edit 3:2 pull-down footage without removing the pull-down first, you need to be particularly careful to match the five-frame pull-down *cadence* at every edit. Edits with broken cadence, such as a repeating or out-of-order frame (for example, A, B, A, B, C, D) can confuse reverse telecine operations. In general, you should avoid editing 29.97 fps pull-down footage. Instead, remove the pull-down of your footage first, edit at 23.98 fps, then reinsert pull-down during output.

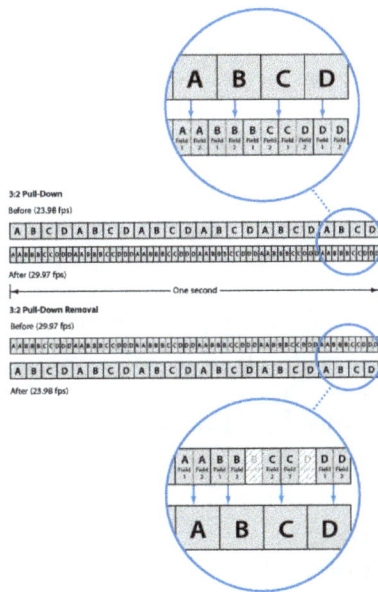

Several NTSC and 1080i60 HD camcorders can record using standard pull-down, though advanced pull-down is usually recommended when recording 24p video. However, for final playback on television or DVD, 3:2 pull-down is generally considered to have the most acceptable quality of motion.

2:3:3:2 Advanced Pull-Down

Camcorders such as the Panasonic AG-DVX100, the Panasonic AG-HVX200, and the Canon XL2 use this method to store 23.98 fps video within interlaced 29.97 fps footage. Video frames 1, 2, 4, and 5 in the pull-down pattern represent film frames A, B, C, and D. Removing advanced pull-down is more efficient than removing standard 2:3:2:3 pull-down because no deinterlacing is required.

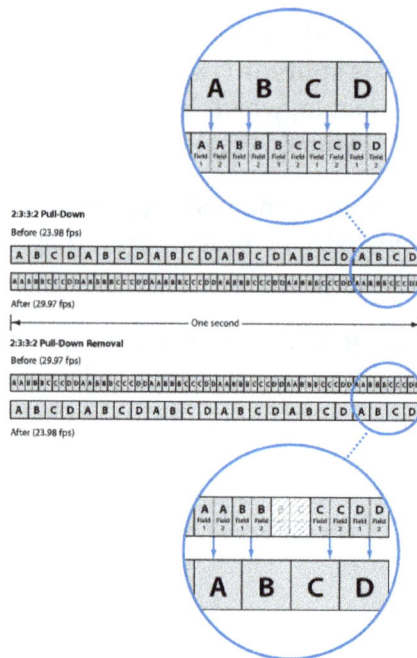

To remove advanced pull-down, video frame 3 in the five-frame pattern is simply discarded during capture or ignored during playback.

Another feature that makes advanced pull-down removal more efficient is the insertion of "flags" in the video signal that can be used by software to automatically detect which frames must be removed. This makes advanced pull-down an automatic process compared to the manual cadence identification usually required to remove 3:2 pull-down.

Despite its efficiencies, the advanced pull-down pattern is not as aesthetically pleasing as 3:2 pull-down. If you plan to finish your project at 23.98 fps, advanced pull-down is usually the best choice. However, if you plan to output your final 24p project to 29.97 fps interlaced video, you may want to add 3:2 pull-down because its pattern is considered to be more visually appealing.

2:2:2:4 Pull-Down

This is an efficient but low-quality playback option used for previewing 23.98 fps footage on an NTSC monitor. Few systems can reverse this kind of pull-down, so you should never record footage with this kind of pull-down. This option is available for situations when processing power is at a premium and your system is unable to generate 2:3:2:3 pull-down or advanced pull-down during playback.

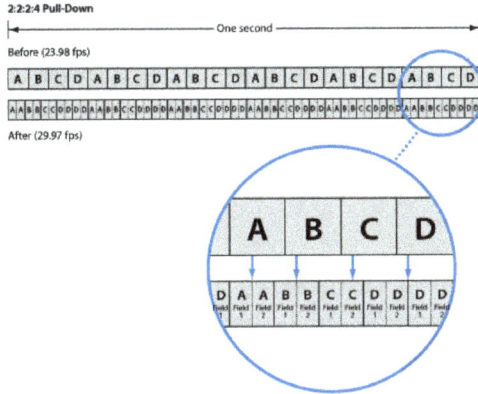

720p DVCPRO HD Duplicate Frames

720p DVCPRO HD camcorders can record 24 fps footage within a 60 fps signal by duplicating frames. The duplicate frames are usually flagged within the DVCPRO HD video signal so applications like Final Cut Pro can automatically remove them. You can also remove duplicate frames using a frame rate converter (such as the DVCPRO HD Frame Rate Converter in Final Cut Pro).

The duplicate frame pattern used in 720p24 footage is similar to the standard 3:2 NTSC telecine pull-down pattern, but there is no interlacing because 720p video is progressively scanned.

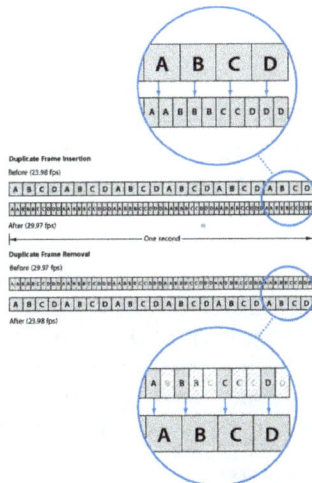

Progressive Segmented Frame Recording

Sony CineAlta cameras can record 23.98 or true 24 fps Progressive segmented Frame (PsF) footage on HDCAM or HDCAM SR tape. The camera records at 48 fields per second while each progressive frame is placed on two fields, resulting in 24 fps.

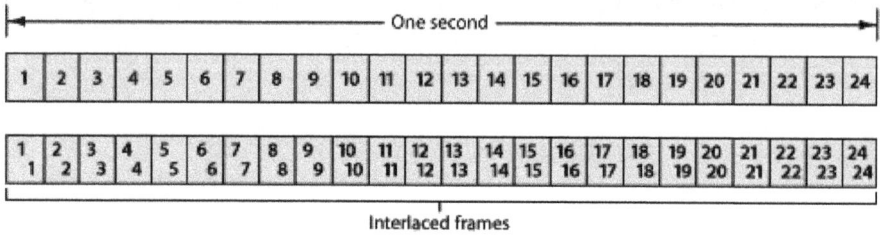

Interlaced frames

24 @ 25

True 24 fps film or video can be transferred to PAL (25 fps) by speeding up the frame rate by 4 percent. For film editing purposes on PAL video, applications like Cinema Tools can slow the 25 fps PAL video back to 24 fps (a process called *conforming*) so that sync is maintained with the original audio.

For showing film-originated movies on PAL video, both film and audio speed are increased by 4 percent. The speed increase is considered acceptable, although the audio must be "pitch shifted" down to match the original. This is the most common method for film-to-PAL transfers.

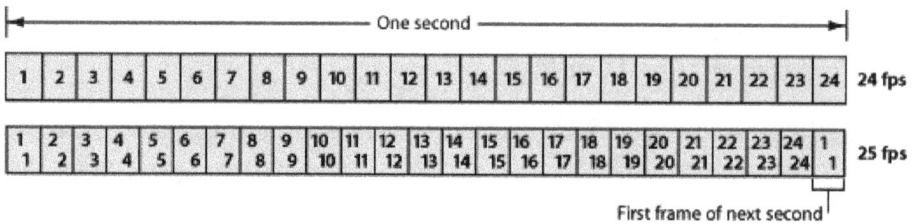

First frame of next second

24 @ 25 Pull-Down

This method does not change the speed of the original film. Instead, film frames 12 and 24 are pulled down for a duration of three fields instead of two, creating a subtle stutter each half second. This pattern is technically described as *2:2:2:2:2:2:2:2:2:2:2:3 pull-down*.

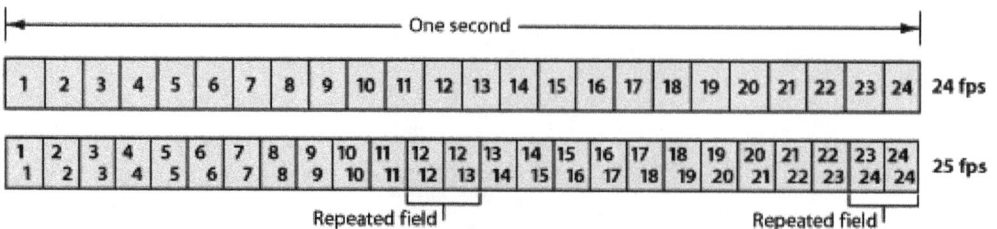

Repeated field   Repeated field

24 @ 25 Repeat

This method simply repeats every 24th frame once to fit 24 fps footage into 25 fps. This causes a noticeable stutter every second but requires less processing than the 24 @ 25 pull-down pattern because no special interlacing is required. This pull-down pattern is analogous to the NTSC 2:2:2:4 pull-down pattern in the sense that it requires the least amount of processing power but results in the most noticeable stutter. You should use this option for preview purposes only and avoid it for final output.

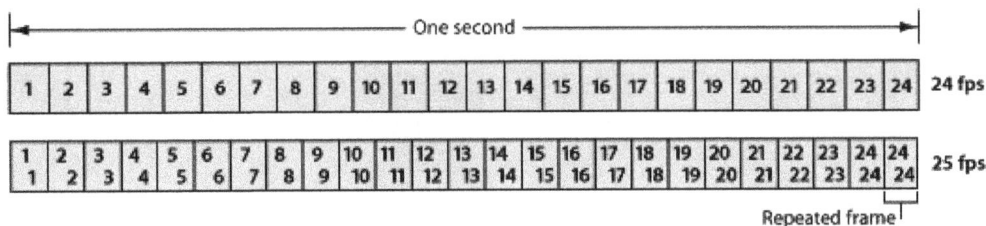

Native 24p

Some video camcorders that record to file-based media can record at 24 or 23.98 fps. For example, the Panasonic AG-HVX200 can record 23.98 fps footage directly. Digital cinema cameras such as the Panavision Genesis, the Dalsa Origin, and the RED ONE can record natively at 24 fps. Of course, film is also recorded at 24 fps.

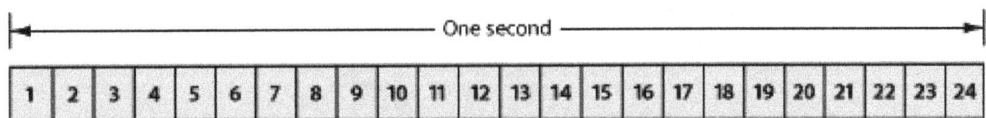

Frame Rate Differences

The most complex part of telecine is the synchronization of the mechanical film motion and the electronic video signal. Every time the video (tele) part of the telecine samples the light electronically, the film (cine) part of the telecine must have a frame in perfect registration and ready to photograph. This is relatively easy when the film is photographed at the same frame rate as the video camera will sample, but when this is not true, a sophisticated procedure is required to change frame rate.

To avoid the synchronization issues, higher end establishments now use a scanning system rather than just a telecine system. This allows them to scan a distinct frame of digital video for each frame of film, providing higher quality than a telecine system would be able to achieve. Normally, best results are then achieved by using a smoothing (interpolating algorithm) rather than a frame duplication algorithm (such as 3:2 pulldown, etc.) to adjust for speed differences between the film and video frame rate.

Similar issues occur when using vertical synchronization to prevent screen tearing, which is a different problem encountered when frame rates mismatch.

2:2 Pulldown

In countries that use the PAL or SECAM video standards, film destined for television is photographed at 25 frames per second. The PAL video standard broadcasts at 25 frames per second, so the transfer from film to video is simple; for every film frame, one video frame is captured.

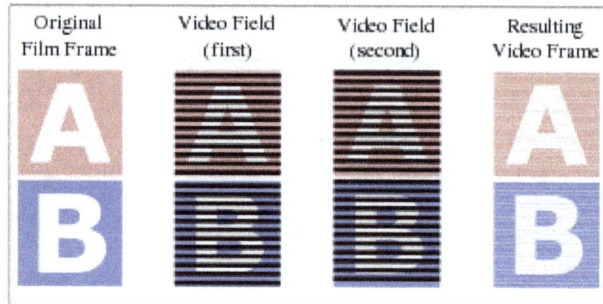

Theatrical features originally photographed at 24 frame/s are shown at 25 frame/s. While this is usually not noticed in the picture (but may be more noticeable during action speed, especially if footage was filmed undercranked), the 4% increase in play-back speed causes a slightly noticeable increase in audio pitch by just over 0.679 semitones, which is sometimes corrected using a pitch shifter, though pitch shifting is a recent innovation and supersedes an alternative method of telecine for 25 frame/s formats.

2:2 pulldown is also used to transfer shows and films, photographed at 30 frames per second, like *Friends* and *Oklahoma!* (1955), to NTSC video, which has 60 Hz scanning rate.

Although the 4% speed increase has been standard since the early days of PAL and SECAM television, recently a new technique has gained popularity, and the resulting speed and pitch of the telecined presentation are identical to that of the original film.

This pulldown method is sometimes used in order to convert 24 frame/s material to 25 frame/s. Usually, this involves a film to PAL transfer without the aforementioned 4% speedup. For film at 24 frame/s, there are 24 frames of film for every 25 frames of PAL video. In order to accommodate this mismatch in frame rate, 24 frames of film have to be distributed over 50 PAL fields. This can be accomplished by inserting a pulldown field every 12 frames, thus effectively spreading 12 frames of film over 25 fields (or "12.5 frames") of PAL video. The method used is 2:2:2:2:2:2:2:2:2:2:2:3 (Euro) pulldown.

This method was born out of a frustration with the faster, higher pitched soundtracks that traditionally accompanied films transferred for PAL and SECAM audiences. A few motion pictures are beginning to be telecined this way. It is particularly suited for films where the soundtrack is of special importance.

When a TV station in an NTSC region airs a film or show that uses a PAL printing/version, but is being broadcast in the NTSC format, sometimes they do not perform the proper PAL to NTSC pulldown conversion or it is done improperly. This causes the program to be sped-up slightly and/or sound higher-pitched, due to the faster rate of the PAL 576 lines/50 Hz vs the NTSC 480 lines/60 Hz format.

Some DVD releases of TV shows/movies, such as The Archie Show, of which the episodes were broadcast in NTSC format, but the masters were either lost or damaged beyond usage, uses PAL 576/50 printings with 3:2 pulldown having not been performed properly, resulting in slightly faster film speed and higher pitch. Though some scenes in a few episodes use NTSC footage due to the PAL masters being missing, and the footage being damaged beyond use, causing varying shifts in speed, pitch, and film quality.

2:3 Pulldown

In the United States and other countries where television uses the 59.94 Hz vertical scanning frequency, video is broadcast at 29.97 frame/s. For the film's motion to be accurately rendered on the video signal, a telecine must use a technique called the **2:3** pulldown, also known as 3:2 pulldown, to convert from 24 to 29.97 frame/s.

The term "pulldown" comes from the mechanical process of "pulling" (physically moving) the film downward within the film portion of the transport mechanism, to advance it from one frame to the next at a repetitive rate (nominally 24 frames/s). This is accomplished in two steps. The first step is to slow down the film motion by 1/1000 to 23.976 frames/s. The difference in speed is imperceptible to the viewer. For a two-hour film, play time is extended by 7.2 seconds.

The second step of the 2:3 pulldown is distributing cinema frames into video fields. At 23.976 frame/s, there are four frames of film for every five frames of 29.97 frame/s video:

$$\frac{23.976}{29.97} = \frac{4}{5}$$

These four frames are "stretched" into five by exploiting the interlaced nature of 60 Hz video. For every frame, there are actually two incomplete images or *fields*, one for the odd-numbered lines of the image, and one for the even-numbered lines. There are, therefore, ten fields for every four film frames, which are called *A*, *B*, *C*, and *D*. The telecine alternately places A frame across two fields, B frame across three fields, C frame across two fields and D frame across three fields. This can be written as A-A-B-B-B-C-C-D-D-D or 2-3-2-3 or simply 2-3. The cycle repeats itself completely after four film frames have been exposed:

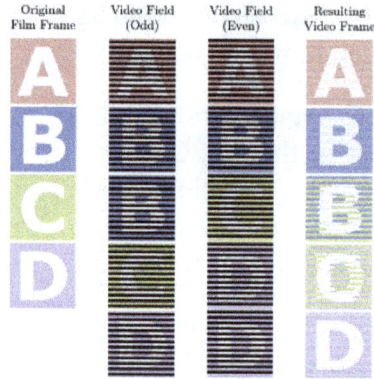

A 3:2 pattern is identical to the one shown above except that it is shifted by one frame. For instance, a cycle that starts with film frame B yields a 3:2 pattern: B-B-B-C-C-D-D-D-A-A or 3-2-3-2 or simply 3-2. In other words, there is no difference between the 2-3 and 3-2 patterns. In fact, the "3-2" notation is misleading because according to SMPTE standards for every four-frame film sequence the first frame is scanned twice, not three times.

The above method is a "classic" 2:3, which was used before frame buffers allowed for holding more than one frame. The preferred method for doing a 2:3 creates only one dirty frame in every five (i.e. 3:3:2:2 or 2:3:3:2 or 2:2:3:3); while this method has slightly more judder, it allows for easier upconversion (the dirty frame can be dropped without losing information) and a better overall compression when encoding. The 2:3:3:2 pattern is supported by the Panasonic DVX-100B video camera under the name "Advanced Pulldown". Note that just fields are displayed—no frames hence no dirty frames—in interlaced display such as on a CRT. Dirty frames may appear in other methods of displaying the interlaced video.

Other Pulldown Patterns

Similar techniques must be used for films shot at "silent speeds" of less than 24 frame/s, which includes home movie formats (the standard for Standard 8 mm film was 16 fps, and 18 fps for Super 8 mm film) as well as silent film (which in 35 mm format usually was 16 fps, 12 fps, or even lower).

- 16 frame/s (actually 15.985) to NTSC 30 frame/s (actually 29.97): pulldown should be 3:4:4:4.

- 16 frame/s to PAL 25: pulldown should be 3:3:3:3:3:3:3:3:4 (a better choice would be to run the film at 16.67 frame/s, simplifying pulldown to 3:3).

- 18 frame/s (actually 17.982) to NTSC 30: pulldown should be 3:3:4.

- 20 frame/s (actually 19.980) to NTSC 30: pulldown should be 3:3.

- 27.5 frame/s to NTSC 30: pulldown should be 3:2:2:2:2.

- 27.5 frame/s to PAL 25: pulldown should be 1:2:2:2:2.

Also, other patterns have been described that refer to the progressive frame rate conversion required to display 24 frame/s video (e.g., from a DVD player) on a progressive display (e.g., LCD or plasma):

- 24 frame/s to 96 frame/s (4x frame repetition): pulldown is 4:4.

- 24 frame/s to 120 frame/s (5x frame repetition): pulldown is 5:5.

- 24 frame/s to 120 frame/s (3:2 pulldown followed by 2x deinterlacing): pulldown is 6:4.

Telecine Judder

The "2:3 pulldown" telecine process creates a slight error in the video signal compared to the original film frames that can be seen in the above image. This is one reason why films viewed on typical NTSC home equipment may not appear as smooth as when viewed in a cinema and PAL home equipments. The phenomenon is particularly apparent during slow, steady camera movements which appear slightly jerky when telecined. This process is commonly referred to as telecine judder.

PAL material in which 2:2:2:2:2:2:2:2:2:2:2:3 (Euro) pulldown has been applied suffers from a similar lack of smoothness, though this effect is not usually called "telecine judder". Effectively, every 12th film frame is displayed for the duration of three PAL fields (60 milliseconds), whereas the other 11 frames are each displayed for the duration of two PAL fields (40 milliseconds). This causes a slight "hiccup" in the video about twice a second. It is increasingly being referred to as Euro pulldown as it largely affects European territories.

Reverse Telecine (a.k.a. Inverse Telecine (IVTC), Reverse Pulldown)

Some DVD players, line doublers, and personal video recorders are designed to detect and remove 2:3 pulldown from telecined video sources, thereby reconstructing the original 24 frame/s film frames. This technique is known as "reverse" or "inverse" telecine. Benefits of reverse telecine include high-quality non-interlaced display on compatible display devices and the elimination of redundant data for compression purposes.

Reverse telecine is crucial when acquiring film material into a digital non-linear editing system such as Lightworks, Sony Vegas Pro, Avid, or Final Cut Pro, since these machines produce negative cut lists which refer to specific frames in the original film material. When video from a telecine is ingested into these systems, the operator usually has available a "telecine trace," in the form of a text file, which gives the correspondence between the video material and film original. Alternatively, the video transfer may include telecine sequence markers "burned in" to the video image along with other identifying information such as time code.

It is also possible, but more difficult, to perform reverse telecine without prior knowledge of where each field of video lies in the 2:3 pulldown pattern. This is the task faced by most consumer equipment such as line doublers and personal video recorders. Ideally, only a single field needs to be identified, the rest following the pattern in lock-step. However, the 2:3 pulldown pattern does not necessarily remain consistent throughout an entire program. Edits performed on film material after it undergoes 2:3 pulldown can introduce "jumps" in the pattern if care is not taken to preserve the original frame sequence (this often happens during the editing of television shows and commercials in NTSC format). Most reverse telecine algorithms attempt to follow the 2:3 pattern using image analysis techniques, e.g. by searching for repeated fields.

Algorithms that perform 2:3 pulldown removal also usually perform the task of deinterlacing. It is possible to algorithmically determine whether video contains a 2:3 pulldown pattern or not, and selectively do either reverse telecine (in the case of filmsourced video) or deinterlacing (in the case of native video sources).

Telecine Hardware

Flying Spot Scanner

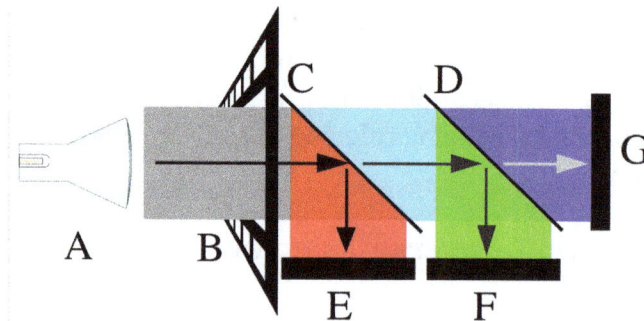

The parts of a flying spot scanner: (A) Cathode-ray tube (CRT); (B) film plane; (C) & (D) dichroic mirrors; (E), (F) & (G) red-, green- and blue-sensitive photomultipliers.

In the United Kingdom, Rank Precision Industries was experimenting with the flying-spot scanner (FSS), which inverted the cathode ray tube (CRT) concept of scanning using a television screen. The CRT emits a pixel-sized electron beam which is converted to a photon beam through the phosphors coating the envelope. This dot of light is then focused by a lens onto the film's emulsion, and finally collected by a pickup device. In 1950 the first Rank flying spot monochrome telecine was installed at the BBC's Lime Grove Studios. The advantage of the FSS is that colour analysis is done after scanning, so there can be no registration errors as can be produced by vidicon tubes where scanning is done after colour separation—it also allows simpler dichroics to be used.

In a flying spot scanner (FSS) or cathode-ray tube (CRT) telecine, a pixel-sized light beam is projected through exposed and developed motion picture film (either negative or positive) and collected by a special type of photo-electric cell known as a photomultiplier which converts the light into an electrical signal. The beam of light "scans" across the film image from left to right to record the horizontal frame information. Vertical scanning of the frame is then accomplished by moving the film past the CRT beam. In a colour telecine the light from the CRT passes through the film and is separated by dichroic mirrors and filters into red, green and blue bands. Photomultiplier tubes or avalanche photodiodes convert the light into separate red, green and blue electrical signals for further electronic processing. This can be accomplished in "real time", 24 frames per second (or in some cases faster). Rank Precision-Cintel introduced the "Mark" series of FSS telecines. During this time advances were also made in CRTs, with increased light output producing a better signal-to-noise ratio and so allowing negative film to be used.

The problem with flying-spot scanners was the difference in frequencies between television field rates and film frame rates. This was solved first by the Mk. I Polygonal Prism system, which was optically synchronised to the television frame rate by the rotating prism and could be run at any frame rate. This was replaced by the Mk. II Twin Lens, and then around 1975, by the Mk. III Hopping Patch (jump scan). The Mk. III series progressed from the original "jump scan" interlace scan to the Mk. IIIB which used a progressive scan and included a digital scan converter (Digiscan) to output interlaced video. The Mk. IIIC was the most popular of the series and used a next generation Digiscan plus other improvements.

The "Mark" series was then replaced by the Ursa (1989), the first in their line of telecines capable of producing digital data in 4:2:2 color space. The Ursa Gold (1993) stepped this up to 4:4:4 and then the Ursa Diamond (1997), which incorporated many third-party improvements on the Ursa system. Cintel's C-Reality and ITK's Millennium flying-spot scanner are able to do both HD and Data.

Line Array CCD

The Robert Bosch GmbH, Fernseh Div., which later became BTS inc. - Philips Digital Video Systems, Thomson's Grass Valley and now is DFT Digital Film Technology introduced the world's first CCD telecine (1979), the FDL-60. The FDL-60 designed and made in Darmstadt West Germany, was the first all solid state telecine.

Rank Cintel (ADS telecine 1982) and Marconi Company (1985) both made CCD Telecines for a short time. The Marconi model B3410 telecine sold 84 units over a three-year period, and a former Marconi technician still maintains them.

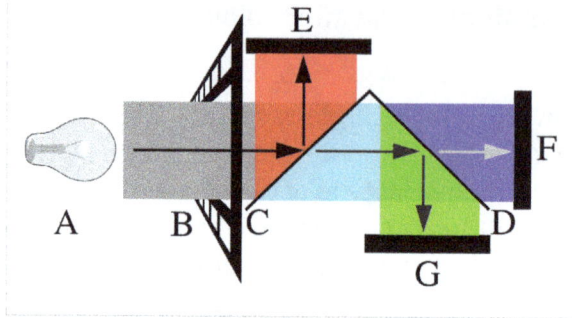

The parts of a CCD scanner: (A) Xenon bulb; (B) film plane; (C) & (D) prisms and/or dichroic mirrors; (E),(F) & (G) red-, green- and blue-sensitive CCDs.

In a charge-coupled device Line Array CCD telecine, a "white" light is shone through the exposed film image into a prism, which separates out the image into the three primary colors, red, green and blue. Each beam of colored light is then projected at a different CCD, one for each color. The CCD converts the light into electrical impulses which the telecine electronics modulate into a video signal which can then be recorded onto video tape or broadcast.

Telecine Shadow system, Denmark

Philips-BTS eventually evolved the FDL 60 into the FDL 90 (1989)/ Quadra (1993). In 1996 Philips, working with Kodak, introduced the Spirit DataCine (SDC 2000), which was capable of scanning the film image at HDTV resolutions and approaching 2K (1920 Luminance and 960 Chrominace RGB) × 1556 RGB. With the data option the Spirit DataCine can be used as a motion picture film scanner outputting 2K DPX data files as 2048 × 1556 RGB. In 2000 Philips introduced the Shadow Telecine(STE), a low cost version of the Spirit with no Kodak parts. The Spirit DataCine, Cintel's C-Reality and ITK's Millennium opened the door to the technology of digital intermediates, wherein telecine tools were not just used for video outputs, but could now be used for high-resolution data that would later be recorded back out to film. The DFT Digital Film Technology, formerly Grass Valley Spirit 4K/2K/HD (2004) replaced the Spirit 1 Datacine and uses both 2K and 4K line array CCDs. (Note: the SDC-2000 did not use a color prisms and/or dichroic mirrors.) DFT revealed its new scanner at the 2009 NAB Show, Scanity. The Scanity uses Time Delay Integration (TDI) sensor technology for extremely fast and sensitive film scans. High speed scanning 15 frame/s @ 4K; 25 frame/s @ 2K; 44 frame/s @ 1K.

Pulsed LED/triggered Three CCD Camera System

With the manufacturing of new high power LEDs, came pulsed LED/triggered three CCD camera systems. Flashing the LED light source for a very short time span gives the full frame CCD camera a stop action of the film, allowing continuous film motion. With CCD video cameras that have a trigger input, the camera now can be electronically synced up to the film transport framing. There are now a number of retail and home made Pulsed LED/triggered camera systems. An array of high-power multiple red, green and blue LEDs is pulsed just as the film frame is positioned in front of the optical lens. The camera sends the single, non-interlaced image of the film frame to a digital frame store, where the electronic picture is clocked out at the selected TV frame rate for PAL or NTSC or other standard. More advanced systems replace the sprocket wheel with laser or camera based perf detection and image stabilization system.

Digital Intermediate Systems and Virtual Telecines

Telecine technology is increasingly merging with that of motion picture film scanners; high-resolution telecines, such as those mentioned above, can be regarded as film scanners that operate in real time.

As digital intermediate post-production becomes more common, the need to combine the traditional telecine functions of input devices, standards converters, and colour grading systems is becoming less important as the post-production chain changes to tapeless and filmless operation.

However, the parts of the workflow associated with telecines still remain, and are being pushed to the end, rather than the beginning, of the post-production chain, in the form of real-time digital grading systems and digital intermediate mastering systems, increasingly running in software on commodity computer systems. These are sometimes called virtual telecine systems.

Video Cameras that Produce Telecined Video, and "Film Look"

Some video cameras and consumer camcorders are able to record in progressive "24 frames/s" or "23.976 frames/s". Such a video has cinema-like motion characteristics and is the major component of so-called "film look" or "movie look".

For most "24 frames/s" cameras, the virtual 2:3 pulldown process is happening inside the camera. Although the camera is capturing a progressive frame at the CCD, just like a film camera, it is then imposing an interlacing on the image to record it to tape so that it can be played back on any standard television. Not every camera handles "24 frames/s" this way, but the majority of them do.

Cameras that record 25 frames/s (PAL) or 29.97 frames/s (NTSC) do not need to employ 2:3 pulldown, because every progressive frame occupies exactly two video fields. In the video industry, this type of encoding is called Progressive Segmented Frame (PsF). PsF is conceptually identical to 2:2 pulldown, only there is no film original to transfer from.

Digital Television and High Definition

Digital television and high definition standards provide several methods for encoding film material. Fifty field/s formats such as 576i50 and 1080i50 can accommodate film content using a 4% speed-up like PAL. 59.94 field/s interlaced formats such as 480i60 and 1080i60 use the same 2:3 pulldown technique as NTSC. In 59.94 frame/s progressive formats such as 480p60 and 720p60, entire frames (rather than fields) are repeated in a 2:3 pattern, accomplishing the frame rate conversion without interlacing and its associated artifacts. Other formats such as 1080p24 can decode film material at its native rate of 24 or 23.976 frame/s.

All of these coding methods are in use to some extent. In PAL countries, 25 frame/s formats remain the norm. In NTSC countries, most digital broadcasts of 24 frame/s progressive material, both standard and high definition, continue to use interlaced formats with 2:3 pull-down, even though ATSC allows native 24 and 23.976 frame/s progressive formats which offer the greatest image quality and coding efficiency, and are widely used in motion picture and high definition video production. Nowadays, most HDTV vendors sell LCD televisions in NTSC/ATSC countries capable of 120 Hz or 240 Hz refresh rates and plasma sets capable of 48, 72, or 96 Hz refresh. When combined with a 1080p24-capable source (such as most Blu-ray Disc players), some of these sets are able to display film-based content using a pulldown scheme of whole multiples of 24, thereby avoiding the problems associated with 2:3 pulldown or the 4% speed-up used in PAL countries. For example, a 1080p 120 Hz set which accepts a 1080p24 input can achieve 5:5 pulldown by simply repeating each frame five times and thus not exhibit picture artifacts associated with telecine judder.

Gate Weave

Gate weave, known in this context as "telecine weave" or "telecine wobble", caused by the movement of the film in the telecine machine gate, is a characteristic artifact of real-time telecine scanning. Numerous techniques have been tried to minimize gate weave, using both improvements in mechanical film handling and electronic post-processing. Line-scan telecines are less vulnerable to frame-to-frame judder than machines with conventional film gates, and non-real-time machines are also less vulnerable to gate weave than real-time machines. Some gate weave is inherent in film cinematography, as it introduced by the film handling within the original film camera: modern digital image stabilization techniques can remove both this and telecine/scanner gate weave.

Soft and Hard Telecine

On DVDs, telecined material may be either hard telecined, or soft telecined. In the hard-telecined case, video is stored on the DVD at the playback framerate (29.97

frame/s for NTSC, 25 frame/s for PAL), using the telecined frames as shown above. In the soft-telecined case, the material is stored on the DVD at the film rate (24 or 23.976 frames/s) in the original progressive format, with special flags inserted into the MPEG-2 video stream that instruct the DVD player to repeat certain fields so as to accomplish the required pulldown during playback. Progressive scan DVD players additionally offer output at 480p by using these flags to duplicate frames rather than fields.

NTSC DVDs are often soft telecined, although lower-quality hard-telecined DVDs exist. In the case of PAL DVDs using 2:2 pulldown, the difference between soft and hard telecine vanishes, and the two may be regarded as equal. In the case of PAL DVDs using 2:3 pulldown, either soft or hard telecining may be applied.

Blu-ray offers native 24 frame/s support, allowing 5:5 cadence on most modern televisions.

Uses

Telecine is performed in a color suite. The most challenging part in telecine is in the synchronization of the film motion with the electronic video signal. It is easy to perform when the film is at the same frame rate, but if that is not the case, a complex process is required for changing the frame rate to establish the synchronization. The basic idea behind telecine is to capture each and every frame and store them. For this, the movie is to be recorded at a frame rate of 16/18 fps. However the normal video speed is 25-30 fps and is thus adjusted with the help of a procedure called the pulldown technique. Telecine can also be reversed using a process called inverse-telecine.

Telecine helps film producers, film distributors and others in releasing their product on video and allows video production devices to complete the filmmaking projects. Images can be optically enlarged up to 50 percent without distortion with the help of telecine.

Compared to scanners, telecine images are less stable and lower in quality. Telecine images are also splice intolerant and susceptible to binding. Telecine also does not have the capability to detect and eliminate dust.

Shortcomings

Compared to scanners, telecine images are less stable and lower in quality. Telecine images are also splice intolerant and susceptible to binding. Telecine also does not have the capability to detect and eliminate dust.

Video Denoising

Recently, as the rapid development of digital imaging technology, digital imaging devices have been widely applied in many fields, including computational photography, security monitoring, robot navigation, and military reconnaissance. However, video signals are often contaminated by all kinds of noise during acquisition and transmission, such as optical noise, component noise, sensor noise, and circuit noise. The noise in video signals not only damages the original information and results in unpleasant visual effect, but also affects the effectiveness of further coding or processing such as feature extraction, object detection, motion tracking, and pattern recognition. So, noise reduction in contaminated video sequences should be implemented.

Many video denoising methods have been proposed in the past decade, most of which perform in the spatial domain, temporal domain, or their combination. Methods in spatial domain often produce limited results because they do not take advantage of spatiotemporal correlations of neighboring frames. Methods in temporal domain consider the correlations of neighboring frames, but they are only appropriate for still video. Additionally, the results have artifacts or smear phenomenon when objects motion exist. By combining the spatial domain with temporal domain, impressive results can be produced. However, these methods generally require a huge amount of computation. With the emergence of new multiresolution tools, such as the wavelet transform, video denoising methods performing in transform domain were proposed continually. Now, the transform domain techniques in general, especially the wavelet-based video denoising methods, have been shown to outperform these spatiotemporal video denoising methods. Moreover, methods that combine spatiotemporal domain and transform domain were also proposed, which could produce perfect denoising effect. Similarly, this kind of methods also require huge amount of computation.

However, although video denoising technology has made great progress, most of these methods are unable to obtain ideal effect for large noisy video sequences in low light, which is urgently needed in many fields, especially in the security monitoring field. In this field, the monitoring devices are fixed in some places in general, so the captured video sequences have fixed background. In practical applications, it often requires to see the characteristic both of still and moving objects in the video sequences clearly. This requirement can be met easily in the day time. However, in the night time, because of the low light condition, captured video sequences are contaminated by noise badly. To some extent, existing video denoising methods can reduce the noise of contaminated video sequences, but this is far from enough to meet the requirement.

Video denoising is the process of removing noise from a video signal. Video denoising methods can be divided into:

- Spatial video denoising methods, where image noise reduction is applied to each frame individually.

- Temporal video denoising methods, where noise between frames is reduced. Motion compensation may be used to avoid ghosting artifacts when blending together pixels from several frames.

- Spatial-temporal video denoising methods use a combination of spatial and temporal denoising. This is often referred to as 3D denoising.

It is done in two areas:

They are chroma and luminance, chroma noise is where one see color fluctuations and luminance is where one see light/dark fluctuations. Generally, the luminance noise looks more like film grain while chroma noise looks more unnatural or digital like.

Video denoising methods are designed and tuned for specific types of noise. Typical video noise types are following:

- Analog noise
  - Radio channel artifacts
    o High frequency interference (dots, short horizontal color lines, etc.)
    o Brightness and color channel interference (problems with antenna)
    o Video reduplication – false contouring appearance
  - VHS artifacts
    o Color-specific degradation
    o Brightness and color channel interference (specific type for VHS)
    o Chaotic line shift at the end of frame (lines resync signal misalignment)
    o Wide horizontal noise strips (old VHS or obstruction of magnetic heads)
  - Film artifacts
    o Dust, dirt, spray
    o Scratches
    o Curling (emulsion exfoliation)
    o Fingerprints
- Digital noise
  - Blocking – low bitrate artifacts
  - Ringing – low and medium bitrates artifact especially on animated cartoons
  - Blocks (slices) damage in case of losses in digital transmission channel or disk injury (scratches on DVD)

Different suppression methods are used to remove all these artifacts from video.

Noise Reduction in Images

Noise is a random variation of image Intensity and visible as a part of grains in the image. It may cause to arise in the image as effects of basic physics-like photon nature of light or thermal energy of heat inside the image sensors. may produce at the time of capturing or image transmissio Noise means, the pixels in the image show different intensity values instead of true pixel values that are obtained from image. Noise removal algorithm is the process of removing or reducing the noise from the image. The noise removal algorithms reduce or remove the visibility of noise by smoothing the entire image leaving areas near con ast boundaries. But these methods can obscure fine, low contrast details. The common types of noise that arises in the image are: a) Impulse noise, b) Additive noise   c) Multiplicative noise. Different noises have their own characteristics which make them distinguishable from others. Image noise can also originated in film grain and in the unavoidable shot noise of an ideal photon detector. Image noise is an undesirable by-product of image captured.

Various Sources of Noise in Images

Noise is introduced in the image at the time of image acquisition or transmission. Different factors may be responsible for introduction of noise in the image. The number of pixels corrupted in the image will decide the quantification of the noise. The principal sources of noise in the digital image are: a) The imaging sensor may be affected by environmental conditions during image acquisition. b)Insufficient Light levels and sensor temperature may introduce the noise in the image. c) Interference in the transmission channel may also corrupt the image. d) If dust particles are present on the scanner screen, they can also introduce noise in the image.

Types of Noise

Noise to be any degradation in the image signal caused by external disturbance.If an image is being sent electronically from one place to another via satellite or wireless transmission or through networked cables, we may expect errors to occur in the image signal. These errors will appear on the image output in different ways depending on the type of disturbance in the signal. Usually we know what type of errors to expect and the type of noise on the image; hence we investigate some of the standard noise for eliminating or reducing noise in color image. Image Noise is classified as Am lifier noise (Gaussian noise), Salt-and-pepper noise (Impu e noise), Shot noise, Quantization noise (uniform noise), Film grain, on-isotropic noise, Speckle noise (Multiplicative noise) and Periodic noise.

Images taken with both digital cameras and conventional film cameras will pick up noise from a variety of sources. Further use of these images will often require that the noise be (partially) removed – for aesthetic purposes as in artistic work or marketing, or for practical purposes such as computer vision.

Types

In salt and pepper noise (sparse light and dark disturbances), pixels in the image are very different in color or intensity from their surrounding pixels; the defining characteristic is that the value of a noisy pixel bears no relation to the color of surrounding pixels. Generally this type of noise will only affect a small number of image pixels. When viewed, the image contains dark and white dots, hence the term salt and pepper noise. Typical sources include flecks of dust inside the camera and overheated or faulty CCD elements.

In Gaussian noise, each pixel in the image will be changed from its original value by a (usually) small amount. A histogram, a plot of the amount of distortion of a pixel value against the frequency with which it occurs, shows a normal distribution of noise. While other distributions are possible, the Gaussian (normal) distribution is usually a good model, due to the central limit theorem that says that the sum of different noises tends to approach a Gaussian distribution.

In either case, the noise at different pixels can be either correlated or uncorrelated; in many cases, noise values at different pixels are modeled as being independent and identically distributed, and hence uncorrelated.

Removal

Tradeoffs

In selecting a noise reduction algorithm, one must weigh several factors:

- the available computer power and time available: a digital camera must apply noise reduction in a fraction of a second using a tiny onboard CPU, while a desktop computer has much more power and time

- whether sacrificing some real detail is acceptable if it allows more noise to be removed (how aggressively to decide whether variations in the image are noise or not)

- the characteristics of the noise and the detail in the image, to better make those decisions

Chroma and Luminance Noise Separation

In real-world photographs, the highest spatial-frequency detail consists mostly of variations in brightness ("luminance detail") rather than variations in hue ("chroma detail"). Since any noise reduction algorithm should attempt to remove noise without sacrificing real detail from the scene photographed, one risks a greater loss of detail from luminance noise reduction than chroma noise reduction simply because most scenes have little high frequency chroma detail to begin with. In addition, most people find chroma noise in images more objectionable than luminance noise; the colored blobs are considered "digital-looking" and unnatural, compared to the grainy appearance of luminance noise that some compare to film grain. For these two reasons, most photographic noise reduction algorithms split the image detail into chroma and luminance components and apply more noise reduction to the former.

Most dedicated noise-reduction computer software allows the user to control chroma and luminance noise reduction separately.

Linear Smoothing Filters

One method to remove noise is by convolving the original image with a mask that represents a low-pass filter or smoothing operation. For example, the Gaussian mask comprises elements determined by a Gaussian function. This convolution brings the value of each pixel into closer harmony with the values of its neighbors. In general, a smoothing filter sets each pixel to the average value, or a weighted average, of itself and its nearby neighbors; the Gaussian filter is just one possible set of weights.

Smoothing filters tend to blur an image, because pixel intensity values that are significantly higher or lower than the surrounding neighborhood would "smear" across the area. Because of this blurring, linear filters are seldom used in practice for noise reduction; they are, however, often used as the basis for nonlinear noise reduction filters.

Anisotropic Diffusion

Another method for removing noise is to evolve the image under a smoothing partial differential equation similar to the heat equation, which is called anisotropic diffusion. With a spatially constant diffusion coefficient, this is equivalent to the heat equation or linear Gaussian filtering, but with a diffusion coefficient designed to detect edges, the noise can be removed without blurring the edges of the image.

Non-local Means

Another approach for removing noise is based on non-local averaging of all the pixels in an image. In particular, the amount of weighting for a pixel is based on the degree of similarity between a small patch centered on that pixel and the small patch centered on the pixel being de-noised.

Nonlinear Filters

A median filter is an example of a non-linear filter and, if properly designed, is very good at preserving image detail. To run a median filter:

1. consider each pixel in the image

2. sort the neighbouring pixels into order based upon their intensities

3. replace the original value of the pixel with the median value from the list

A median filter is a rank-selection (RS) filter, a particularly harsh member of the family of rank-conditioned rank-selection (RCRS) filters; a much milder member of that family, for example one that selects the closest of the neighboring values when a pixel›s value is external in its neighborhood, and leaves it unchanged otherwise, is sometimes preferred, especially in photographic applications.

Median and other RCRS filters are good at removing salt and pepper noise from an image, and also cause relatively little blurring of edges, and hence are often used in computer vision applications.

Wavelet Transform

The main aim of an image denoising algorithm is to achieve both noise reduction and feature preservation. In this context, wavelet-based methods are of particular interest. In the wavelet domain, the noise is uniformly spread throughout coefficients while most of the image information is concentrated in a few large ones. Therefore, the first wavelet-based denoising methods were based on thresholding of detail subbands coefficients. However, most of the wavelet thresholding methods suffer from the drawback that the chosen threshold may not match the specific distribution of signal and noise components at different scales and orientations.

To address these disadvantages, non-linear estimators based on Bayesian theory have been developed. In the Bayesian framework, it has been recognized that a successful denoising algorithm can achieve both noise reduction and feature preservation if it employs an accurate statistical description of the signal and noise components.

Statistical Methods

Statistical methods for image denoising exist as well, though they are infrequently used as they are computationally demanding. For Gaussian noise, one can model the pixels in a greyscale image as auto-normally distributed, where each pixel's "true" greyscale value is normally distributed with mean equal to the average greyscale value of its neighboring pixels and a given variance.

Let δ_i denote the pixels adjacent to the ith pixel. Then the conditional distribution of the greyscale intensity (on a scale) at the ith node is:

$$\mathbb{P}(x(i) = c \mid x(j) \forall j \in \delta i) \propto e^{-\frac{\beta}{2\lambda} \sum_{j \in \delta i} (c - x(j))^2}$$

for a chosen parameter $\beta \geq 0$ and variance λ. One method of denoising that uses the auto-normal model uses the image data as a Bayesian prior and the auto-normal density as a likelihood function, with the resulting posterior distribution offering a mean or mode as a denoised image.

Block-matching Algorithms

A block-matching algorithm can be applied to group similar image fragments into overlapping macroblocks of identical size, stacks of similar macroblocks are then filtered together in the transform domain and each image fragment is finally restored to its original location using a weighted average of the overlapping pixels.

Random Field

Shrinkage fields is a random field-based machine learning technique that brings performance comparable to that of Block-matching and 3D filtering yet requires much lower computational overhead (such that it could be performed directly within embedded systems).

Deep Learning

Various deep learning approaches have been proposed to solve noise reduction and such image restoration tasks. Deep Image Prior is one such technique which makes use of convolutional neural network and is distinct in that it requires no prior training data.

Software Programs

Most general purpose image and photo editing software will have one or more noise reduction functions (median, blur, despeckle, etc.). Special purpose noise reduction software programs include Neat Image, Noiseless, Noiseware, Noise Ninja, G'MIC (through the *-denoise* command), and pnmnlfilt (nonlinear filter) found in the open source Netpbm tools. General purpose image and photo editing software including noise reduction functions include Adobe Photoshop, GIMP, PhotoImpact, Paint Shop Pro, Helicon Filter, and Darktable.

Digital Intermediate

The basic definition first. DI stands for Digital Intermediate. It is a process, not an art or a science. Its that part in a film's production that needs to be done after editing and before printing. It's an intermediate stage.

To explain this slightly better, DI is the process of taking an edited film, in the form of an EDL or a cut list, scanning the OK shots at a very high resolution, working on the shots one at a time, fixing the colours to make the film look interesting, maybe doing effects and blending CG with "normal" footage, and finally making out a negative from which the producer can make multiple prints to show in theatres. This negative that comes out is an "intermediate" and it's done digitally, so the process is called "digital intermediate".

Digital Intermediate Workflow

A typical digital intermediate workflow consists of three steps:

1. A film scanner scans the original film negative frame by frame. A typical scanner, such as the Arriscan, flashes each frame with a red, green, and blue light, and each frame is captured on a sensor as a "raw" file that is uncorrected. Based on an EDL (edit decision list) provided by the editor, the film scanner is capable of identifying and selecting each original roll of film to find the exact start and end frame of each needed shot. The scanning process varies from facility to facility and might offer a variety of image resolutions (2K, 4K, 6K; the higher the value, the sharper the image) and color bit depths (such as 10 bits per color channel). Each scanned frame is then recorded onto a hard drive and is numbered sequentially.

2. The image sequence is conformed and manipulated. The scanned film frames are delivered to the title designer as an image sequence so that titles can be composited over the footage. This is also the appropriate time to perform any necessary special effects or color corrections. Look-up tables (LUTs) are frequently used to make sure that the footage will match both the digital projector and the print film stock of choice. Once all the manipulation is completed, the image sequence needs to be prepared and exported so that it can be printed back onto film.

3. The image sequence is printed back onto film (film-out). This step involves the use of a film printer, which reads the information of each digital frame and uses a laser to engrave it frame by frame onto a film roll.

It often replaces or augments the photochemical timing process and is usually the final creative adjustment to a movie before distribution in theaters. It is distinguished from the telecine process in which film is scanned and color is manipulated early in the process to facilitate editing. However the lines between telecine and DI are continually blurred and are often executed on the same hardware by colorists of the same background. These two steps are typically part of the overall color management process in a motion picture at different points in time. A digital intermediate is also customarily done at higher resolution and with greater color fidelity than telecine transfers.

Although originally used to describe a process that started with film scanning and ended with film recording, digital intermediate is also used to describe color correction and color grading and even final mastering when a digital camera is used as the image source and/or when the final movie is not output to film. This is due to recent advances in digital cinematography and digital projection technologies that strive to match film origination and film projection.

In traditional photochemical film finishing, an intermediate is produced by exposing film to the original camera negative. The intermediate is then used to mass-produce the films that get distributed to theaters. Color grading is done by varying the amount of red, green, and blue light used to expose the intermediate. This seeks to be able to replace or augment the photochemical approach to creating this intermediate.

The digital intermediate process uses digital tools to color grade, which allows for much finer control of individual colors and areas of the image, and allows for the adjustment of image structure (grain, sharpness, etc.). The intermediate for film reproduction can then be produced by means of a film recorder. The physical intermediate film that is a result of the recording process is sometimes also called a digital intermediate, and is usually recorded to internegative (IN) stock, which is inherently finer-grain than camera negative (OCN).

One of the key technical achievements that made the transition to DI possible was the use of the 3D look-up tables (aka "3D LUTs"), which could be used to mimic how the digital image would look once it was printed onto release print stock. This removed a large amount of skilled guesswork from the film-making process, and allowed greater freedom in the colour grading process while reducing risk.

The digital master is often used as a source for a DCI-compliant distribution of the motion picture for digital projection. For archival purposes, the digital master created during the Digital Intermediate process can still be recorded to very stable high dynamic range yellow-cyan-magenta (YCM) separations on black-and-white film with an expected 100-year or longer life. This archival format, long used in the industry prior to the invention of DI, still provides an archival medium that is independent of changes in digital data recording technologies and file formats that might otherwise render digitally archived material unreadable in the long term.

Color Grading

Color grading is the creative process where decisions are made to further enhance or establish a new visual tone to the project through software including: introducing new color themes, re-lighting within a frame, films stock emulations, color gradients and a slew of other choices. Being that this is purely creative, there is no wrong or right only what the DP, director and colorist feel is appropriate for the story. It can be subtle and invisible or over-the-top and uber-stylized. Therein lies the challenge. The challenge of choices. The tools available are so numerous, powerful and often free (Davinci Resolve

Lite!) that you have no excuse not to explore these options further before you embark on the Grading journey.

Color Grading Purpose

Color Grading is a multi-process that can change the visual tone of an entire film. Once your footage is corrected, you can work to change the thematics and aesthetics. Grading is used more as a brush to paint a picture with purpose. These include:

- Shot Matching

- Removing Objects

- Shape Masks

- Cinematic Looks (day-to-night, underwater, flashbacks, etc...)

Color Timing

Color timing is the colorization of film as it is being developed, involving the photo-chemical process in creating colorized prints. Color timing was used extensively with films before the digital age and performed in a laboratory. Color timing is used to manipulate the color and give the scene a consistent look between shots. It has a great effect on filmed images, as it controls the 'look' of the film, with respect to exposure and color balance, as well as scene-to-scene continuity.

The color timer uses a machine known as a 'Hazeltine' which reverses images on the original negatives and displays them on a television-like screen, and then turns dials to assign the image 'printers points' for each of the three primary colors (red, green, blue). These 'points' range from 0 to 50, with about 25 being 'normal,' with higher numbers making the image darker, and lower numbers making the image lighter. Each scene is timed, and the printer's points for each scene are encoded onto a punched paper tape. Once the final print and color options are locked, the film is printed directly off via laser and the traditional color-timing stage is finished. The term 'timer' comes from the days before automated printers when the 'timer' had to determine how long certain portions of be allowed to sit in the developer.

Oh Brother Where Art Thou is a modern update on this concept. It was the first film to really harness the digital color grading process, even before the term "digital color grading" existed. It has since been used on several other films, including Lord of the Rings and Pleasantville.

Telecine

With the advent of television, broadcasters quickly realised the limitations of live television broadcasts and they turned to broadcasting feature films from release prints directly from a telecine. This was before 1956 when Ampex introduced the first Qua-druplex videotape recorder (VTR) VRX-1000. Live television shows could also be re-corded to film and aired at different times in different time zones by filming a video monitor. The heart of this system was the *kinescope*, a device for recording a television broadcast to film.

The early telecine hardware was the "film chain" for broadcasting from film and utilsed a film projector connected to a video camera. As explained by Jay Holben in *American Cinematographer Magazine*, "The telecine didn't truly become a viable post-production tool until it was given the ability to perform colour correction on a video signal."

Today, telecine is synonymous with colour timing as tools and technologies have ad-vanced to make color timing (colour correction) ubiquitous in a video environment.

How Telecine Colouring Works

In a Cathode-ray tube (CRT) system, an electron beam is projected at a phosphor-coat-ed envelope, producing a spot of light the size of a single pixel. This beam is then scanned across a film frame from left to right, capturing the "vertical" frame informa-tion. Horizontal scanning of the frame is then accomplished as the film moves past the CRT's beam. Once this photon beam passes through the film frame, it encounters a series of dichroic mirrors which separate the image into its primary red, green and blue components. From there, each individual beam is reflected onto a photomul-tiplier tube (PMT) where the photons are converted into an electronic signal to be recorded to tape.

In a charge-coupled device (CCD) telecine, a white light is shone through the exposed film image onto a prism, which separates the image into the three primary colors, red, green and blue. Each beam of colored light is then projected at a different CCD, one for each color. The CCD converts the light into an electronic signal, and the telecine elec-tronics modulate these into a video signal that can then be color graded.

Early color correction on Rank Cintel MkIII CRT telecine systems was accomplished by varying the primary gain voltages on each of the three photomultiplier tubes to vary the output of red, green and blue. Further advancements converted much of the color-pro-cessing equipment from analog to digital and then, with the next-generation telecine, the Ursa, the coloring process was completely digital in the 4:2:2 color space. The Ursa Gold brought about color grading in the full 4:4:4 color space.

Color correction control systems started with the Rank Cintel TOPSY (Telecine Op-erations Programming SYstem) in 1978. In 1984 Da Vinci Systems introduced their first color corrector, a computer-controlled interface that would manipulate the color voltages on the Rank Cintel MkIII systems. Since then, technology has improved to give extraordinary power to the digital colorist. Today there are many companies making color correction control interfaces including Da Vinci Systems, Pandora International, Pogle and more.

Some of the main functions of electronic (digital) color grading:

- Reproduce accurately what was shot

- Compensate for variations in the material (i.e., film errors, white balance, varying lighting conditions)

- Optimize transfer for use of special effects

- Establish a desired 'look'

- Enhance and/or alter the mood of a scene — the visual equivalent to the musical accompaniment of a film; compare also film tinting

Note that some of these functions are contrary to others; for example, color grading is often done to ensure that the recorded colors match those of the set design, whereas in music videos, the goal may instead be to establish a stylized look.

Traditionally, color grading was done towards technical goals. For example, in the film Marianne, grading was used so that night scenes could be filmed more cheaply in daylight. Secondary color correction was originally used to establish color continuity, however the trend today is increasingly moving towards creative goals, such as improving the aesthetics of an image, establishing stylized looks, and setting the mood of a scene through color. Due to this trend, some colorists suggest the phrase "color enhancement" over "color correction".

Primary and Secondary Color Grading

Primary color grading affects the whole image by providing control over the color density curves of red, green, blue color channels, across the entire frame. Secondary correction can isolate a range of hue, saturation and brightness values to bring about alterations in hue, saturation and luminance only in that range, allowing the grading of secondary colors, while having a minimal or usually no effect on the remainder of the color spectrum. Using digital grading, objects and color ranges within a scene can be isolated with precision and adjusted. Color tints can be manipulated and visual treatments pushed to extremes not physically possible with laboratory processing. With these advancements, the color correction process has become increasingly similar to well-established digital painting techniques, ushering forth a new era of digital cinematography.

Masks, Mattes, Power Windows

The evolution of digital color grading tools has advanced to the point where the colorist can use geometric shapes (such as mattes or masks in photo software such as Adobe

Photoshop) to isolate color adjustments to specific areas of an image. These tools can highlight a wall in the background and color only that wall, leaving the rest of the frame alone, or color everything but that wall. Subsequent color correctors (typically software-based) have the ability to use spline-based shapes for even greater control over isolating color adjustments. Color keying is also used for isolating areas to adjust.

Inside and outside of area-based isolations, digital filtration can be applied to soften, sharpen or mimic the effects of traditional glass photographic filters in nearly infinite degrees.

Motion Tracking

When trying to isolate a color adjustment on a moving subject, the colorist traditionally would have needed to manually move a mask to follow the subject. In its most simple form, motion tracking software automates this time-consuming process using algorithms to evaluate the motion of a group of pixels. These techniques are generally derived from match moving techniques used in special effects and compositing work.

Digital Intermediate

The evolution of the telecine device into film scanning allowed the digital information gathered from a film negative to be of sufficient resolution to transfer back to film. In the late 1990s, the films *Pleasantville* and *O Brother, Where Art Thou?* advanced the technology to the point that the creation of a digital intermediate was possible, which greatly expanded the capabilities of the digital telecine colorist in a traditionally film-oriented world. Today, many feature films go through the DI process, while manipulation through photochemical processing is decreasing in use.

In Hollywood, *O Brother, Where Art Thou?* was the first film to be wholly digitally graded. The negative was scanned with a Spirit DataCine at 2K resolution, then colors were digitally fine-tuned using a Pandora MegaDef color corrector on a Virtual Data-Cine. The process took several weeks, and the resulting digital master was output to film again with a Kodak laser recorder to create a master internegative.

Modern motion picture processing typically uses both digital cameras and digital projectors; when done correctly, color correction in such a system is a technical function involving the calibration of the different elements of the system, leaving the color grading process entirely to the creation of artistic color effects.

Hardware-based versus Software-based Systems

Hardware-based systems (da Vinci 2K, Pandora International MegaDEF, etc.) have historically offered better performance and a smaller feature set than software-based systems. Their real time performance was optimised to particular resolution and bit depths, as opposed to software platforms using standard computer industry hardware that often trade speed for resolution independence, e.g. Apple's Color (previously Silicon Color Final Touch), ASSIMILATE SCRATCH, Adobe SpeedGrade and SGO Mistika. While hardware-based systems always offer real-time performance, some software-based systems need to render as the complexity of the color grading increases. On the other hand, software-based systems tend to have more features such as spline-based windows/masks and advanced motion tracking.

The line between hardware and software is blurring as many software-based color correctors (e.g. Pablo, Mistika, SCRATCH, Autodesk Lustre, Nucoda Film Master and Filmlight Baselight) use multi processor workstations and a GPU (graphics processing unit) as a means of hardware acceleration. As well, some newer software-based systems use a cluster of multiple parallel GPUs on the one computer system to improve performance at the very high resolutions required for feature film grading. e.g. Blackmagic Designs' DaVinci Resolve. Some color grading software like Synthetic Aperture's Color Finesse runs solely as software and will even run on low-end computer systems.

Hardware

The control panels are placed in a color suite for the colorist to operate.

- For high-end systems many telecines are controlled by a Da Vinci Systems color corrector 2k or 2k Plus, which is also called color grading.

- Other high-end systems are controlled by Pandora Int.'s Pogle, often with either a MegaDEF, Pixi, or Revolution color grading system.

- Additionally, color grading systems require an edit controller. The edit controller controls the telecine and a VTR(s) or other recording/playback devices to ensure frame accurate film frame editing. There are a number of systems which can be used for edit control. Some color grading products such as Pandora Int.'s Pogle have a built in edit controller. Otherwise, a separate device such as Da Vinci Systems' TLC edit controller would be used.

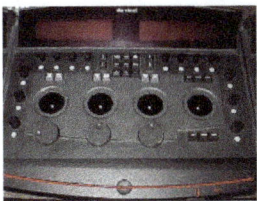

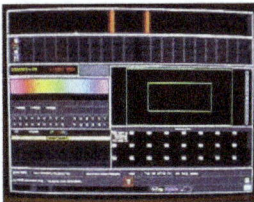

Da Vinci Joy ball con- Da Vinci 2k Display   Pogle control panel and   Pogle SGI monitor
trol panel   soft knob display   display

- Older systems are: Renaissance, Classic analog, Da Vinci Systems's: The Whiz (1982) and 888; The Corporate Communications's System 60XL (1982–1989) and Copernicus-Sunburst; Bosch Fernseh's FRP-60 (1983–1989); Dubner (1978–1985?), Cintel's TOPSY (1978), Amigo (1983), and ARCAS (1992)

systems. All of these older systems work only with standard-definition 525 and 625 video signals, and are considered near obsolete today.

| Pogle control panel joyballs | DFT Scanity | SDC-2000 Spirit Data-Cine Film Deck | Bosch Fernseh FDL 60 Telecine Film Deck and Lens Gate |

Software

The controls are shown on-screen and are sometimes accessed as a plugin to a host application.

- Baselight from FilmLight is used for HD, 2K, 4K and 3D color grading. Grade operations are controlled via Blackboard. Program supports variety of film and video formats and codecs. FilmLight systems utilises cluster and cloud technology in Linux environment.

- Nucoda from Digital Vision provides advance color grading tools working with ACES and HDR at SD to 8K as well as industry leading restoration and image enhancement tools.

- Software like Synthetic Aperture's Color Finesse runs as a plugin in host applications like Apple's Final Cut Pro, Adobe's After Effects and Premiere.

- Da Vinci Systems from Blackmagic Design operates on Mac OS X, Windows 7 Pro and Linux OS utilizing a cluster of multiple parallel GPUs for real time grading of HD, 2K and 4K images in 2D or Stereoscopic 3D.

- SpeedGrade from Adobe Systems released as a part of Creative Suite 6 and Creative Cloud works on Mac and PC. It works on layers interface and the workflow is linked to Premiere Pro and After Effects.

- Magic Bullet Colorista II from Red Giant Software offers multi-step color correction with primary, secondary and master stages inside host applications including Apple's Final Cut Pro, Adobe's After Effects and Premiere.

- The Grading Sweet is a package of specialized Color Grading plugins for Apple's Final Cut Pro.

- Sony Vegas has many built-in filters as well as third-party plugins for color grading.

- Apple Final Cut Studio 2 contains Apple Color which is a dedicated software application for color grading.

- Bones Dailies by Digital Film Technology.

- Other programs have their own color grading options (for example Edius or Blender).

- Autodesk Lustre is a high-end color grading solution. It features GPU acceleration for most functions.

- YUVsoft Color Corrector Adobe After Effects plug-in for Stereo3D color grading.

- Mistika (SGO) is a color grading and online editing system.

- Quantel's Pablo Rio color correction and finishing system is available as software only or in a range of turnkey configurations.

- Assimilate Scratch has advanced color grading and compositing tools and is used for creating digital dailies and for final finishing. It runs in Mac and Windows environments.

- Film Convert is a simple color grading tool that converts digital footage to emulate the look of real film stocks

Motion Interpolation

Motion interpolation is a feature that increases a video's frame rate, usually up to the maximum frame rate of the TV. This has the effect of making movements look smoother, clearer, and more lifelike than what you typically see from movies and TV – pretty similar to the look of movement in soap operas (hence the alternate name of 'soap opera effect'). This feature only matters if you want to be able to make movement in videos look a bit smoother.

The goal of motion interpolation is to give the viewer a more life-like picture. Some viewers, however, think the picture is too lifelike and that motion interpolation makes films on TV look as if they were raw video feeds. (Soap operas have traditionally been recorded on video, not film.)

How it Works

If a television screen has a refresh rate of 120Hz (120 frames per second) but the television is going to display film that was recorded at the standard 24 frames per second, the vendor must figure out a way to fill in an extra 94 frames each second.

One way to do this is to have the television repeat each film frame five times(5x24=120). Another way is to have a computer program in the television digitally analyze concurrent frames and use the data to create intermediary frames. The insertion of these frames is called interpolation and they are what cause the soap opera effect.

Hardware Applications

Displays

Motion interpolation is a common, optional feature of various modern display devices such as HDTVs and video players, aimed at increasing perceived framerate or alleviating display motion blur, a common problem on LCD flat-panel displays.

Difference from Display Framerate

A display's framerate is not always equivalent to that of the content being displayed. In other words, a display capable of or operating at a high framerate does not necessarily mean that it can or must perform motion interpolation (most TVs ship with any such feature enabled by default). For example, a TV running at 120 Hz and displaying 24 FPS content will simply display each content frame for five of the 120 display frames per second. This has no effect on the picture other than eliminating the need for 3:2 pulldown and thus film judder as a matter of course (since 120 is evenly divisible by 24). Eliminating judder results in motion that is less "jumpy" and which matches that of a theater projector. Motion interpolation can be used to reduce judder, but it is not required in order to do so.

Relationship to Advertised Display Framerate

The advertised framerate of a specific display may refer to either the maximum number of content frames which may be displayed per second, or the number of times the display is refreshed in some way, irrespective of content. In the latter case, the actual presence or strength of any motion interpolation option may vary. In addition, the ability of a display to show content at a specific framerate does not mean that display is capable of accepting content running at that rate; most consumer displays above 60 Hz do not accept a higher frequency signal, but rather use the extra frame capability to eliminate judder, reduce ghosting, or create interpolated frames.

As an example, a TV may be advertised as "240 Hz", which would mean one of two things:

1. The TV can natively display 240 frames per second, and perform advanced motion interpolation which inserts between 2 and 8 new frames between existing ones (for content running at 60 FPS to 24 FPS, respectively). For active 3D, this framerate would be halved.

2. The TV is natively only capable of displaying 120 frames per second, and basic motion interpolation which inserts between 1 and 4 new frames between existing ones. Typically the only difference from a "120 Hz" TV in this case is the addition of a strobing backlight, which flickers on and off at 240 Hz, once after every 120 Hz frame. The intent of a strobing backlight is to increase the apparent response rate and thus reduce ghosting, which

results in smoother motion overall. However, this technique has nothing to do with actual framerate. For active 3D, this framerate is halved, and no motion interpolation or pulldown functionality is typically provided. 600 Hz is an oft-advertised figure for plasma TVs, and while technically correct, it only refers to an inter-frame response time of 1.6 milliseconds. This can significantly reduce ghosting and thus improve motion quality, but is unrelated to interpolation and content framerate. There are no consumer films shot at 600 frames per second, nor any TV processors capable of generating 576 interpolated frames per second.

HDTV Implementations

The commercial name given to HDTV motion interpolation technology varies across manufacturers, as does its implementation.

- Hitachi – Reel120

- Insignia – DCM Plus, for Digital Clear Motion 120 Hz, or Insignia Motion 120 Hz

- Kogan.com – MotionMax 100 Hz, 200 Hz

- LG – TruMotion 120 Hz, 240 Hz, 480 Hz, MCI 120

- AOC – Motion Boost 120 Hz

- Bose – VideoWave III 120 Hz (Not named)

- Loewe – Digital Movie Mode (DMM)

- Mitsubishi – Smooth 120 Hz

- Panasonic – Intelligent Frame Creation (IFC) 24p Smooth Film (24p material only)

- Philips – HD Digital Natural Motion, Perfect Motion Rate

- Samsung – Auto Motion Plus 120 Hz, 240 Hz, Clear Motion Rate 100 Hz, 200 Hz, 400 Hz, 500 Hz, 600 Hz, 800 Hz; (PAL video system), Clear Motion Rate 120 Hz, 240 Hz, 480 Hz, 600 Hz, 720 Hz, 960 Hz (NTSC video system)

- Sharp – Fine Motion Enhanced, AquoMotion 240 Hz, AquoMotion Pro

- Sony – MotionFlow 100 Hz, 100 Hz PRO (XBR series, Australia), 120 Hz, 200 Hz, 240 Hz, 400 Hz, 480 Hz, 800 Hz, 960 Hz

- Toshiba – ClearScan 120 Hz, 240 Hz

- Vizio – SmoothMotion

- Sceptre – MEMC (Motion Estimation/Motion Compensation)

- Hisense – Ultra Smooth Motion Rate SMR 120

- Westinghouse – MEMC (Motion Estimation/Motion Compensation)

Software Applications

Video Playback Software

Motion interpolation features are included with several video player applications.

- WinDVD uses Philips' TrimensionDNM for frame interpolation.

- PowerDVD uses TrueTheater Motion for interpolation of DVD and video files to up to 72 frame/s.

- Splash PRO uses Mirillis Motion^2 technology for up to Full HD video interpolation.

- DmitriRender uses GPU-oriented Frame Rate Conversion algorithm with native DXVA support for frame interpolation.

- Bluesky Frame Rate Converter – a DirectShow Filter that can convert the frame rate using AMD Fluid Motion.

- SmoothVideo Project comes integrated by default with MPC-HC in free version; paid version can integrate with more players, including VLC.

Video Editing Software

Some video editing software and plugins offer motion interpolation effects to enhance digitally-slowed video. Adobe After Effects has this in a feature called "Pixel Motion". The effects plugin "Twixtor" is available for most major video editing suites, and offers similar functionality.

Virtual Reality

On October 6, 2016, Oculus VR announced that it would enable the use of motion interpolation on the Oculus Rift virtual reality headset, allowing it to be used on computers whose specifications are not high enough to render to the headset at 90 frames per second.

Side Effects

Visual Artifacts

Motion interpolation on certain brands of TVs is sometimes accompanied by visual anomalies in the picture, described by CNET's David Carnoy as a "little tear or glitch" in the picture, appearing for a fraction of a second. He adds that the effect is most noticeable when the technology suddenly kicks in during a fast camera pan. Television and display manufacturers refer to this phenomenon as a type of digital artifact. Due to the improvement of associated technology over time, such artifacts appear less frequently with modern consumer TVs, though they have yet to be eliminated entirely "the artifacts happens more often when the gap between frames are bigger".

Soap Opera Effect

As a byproduct of the perceived increase in framerate, motion interpolation may introduce a "video" (versus "film") look. This look is commonly referred to as the "soap opera effect" (SOE), in reference to the distinctive appearance of most broadcast television soap operas or pre 2000s multicam sitcoms, which were typically shot using less expensive 60i video rather than film. Many complain that the soap opera effect ruins the theatrical look of cinematic works, by making it appear as if the viewer is either on set or watching a behind the scenes featurette. For this reason, almost all manufacturers have built in an option to turn the feature off or lower the effect strength.

Others appreciate motion interpolation as it reduces motion blur produced by camera pans and shaky cameras and thus yields better clarity of such images. It may also be used to increase the apparent framerate of video game software for a more realistic feel, though the addition of input lag may be an undesired side effect. This "video look" is created deliberately by the VidFIRE technique to restore archive television programs that only survive as film telerecordings. The main differences between an artificially and naturally high framerate (via interpolation versus in-camera), are that the latter is not subject to any of the aforementioned artifacts, contains more accurate (or "true to life") image data, and requires more storage space and bandwidth since frames are not produced in realtime.

Motion Compensation

Motion compensation is the use of the motion estimation information to achieve compression. If you can describe the motion, then you only have to describe the changes that occur after compensating for that motion.

Predictive coding is widely used in video transmission, especially for low bit-rate coding. Typically only some fraction of an image changes from frame to frame allowing straightforward prediction from previous frames. Motion compensation is used as part of the predictive process. If an image sequence shows moving objects, then their motion within the scene can be measured, and the information used to predict the content of frames later in the sequence.

Frame Based Block-matching Motion Compensation

Fixed Size Block-Matching (FSBM)

The technique was originally described by Jain and Jain ; it is easy to implement, and thus widely adopted. Each image frame is divided into a fixed number of usually square blocks. For each block in the frame, a search is made in the reference frame over an area of the image that allows for the maximum translation that the coder can use. The search is for the best matching block, to give the least prediction error, usually minimising either mean square difference, or mean absolute difference which is easier to compute.

Typical block sizes are of the order of 16x16 pixels, and the maximum displacement might be +-64 pixels from a block's original position. Several search strategies are possible, usually using some kind of sampling mechanism, but the most straightforward approach is exhaustive search. This is computationally demanding in terms of data throughput, but algorithmically simple, and relatively easily implemented in hardware.

A good match during the search means that a good prediction can be made, but the improvement in prediction must outweigh the cost of transmitting the motion vector. A good match requires that the whole block has undergone the same translation, and the block should not overlap objects in the image that have different degrees of motion, including the background.

The choice of block-size to use for motion compensation is always a compromise, smaller and more numerous blocks can better represent complex motion than fewer large ones. This reduces the work and transmission costs of subsequent correction stages but with greater cost for the motion information itself. The problem has been investigated by Ribas-Corbera and Neuhoff and they conclude that the choice of block-size can be affected not only by motion vector accuracy but also by other scene characteristics such as texture and inter-frame noise.

These motion vectors are subsequently variable length coded using a differential 2D prediction mechanism.

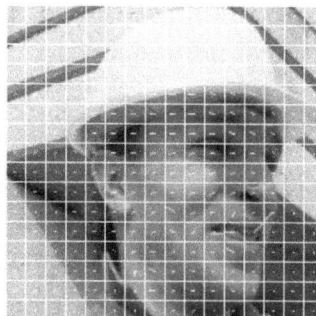

FSBM example.

It is well known that the motion vectors resulting from FSBM are well correlated. Vector information can be coded differentially using variable length codes. This is done in a number of codecs (e.g.. ITU-T H.263) and it is proposed for the MPEG-4 video standard.

An example of the block structure generated is shown in figure (a frame from the MPEG-4 test sequence known as "Foreman"). It can be noticed that the stationary background is represented by large numbers of blocks with very similar motion vectors (represented by the short lines starting from each block centre).

Variable Size Block-Matching (VSBM)

Proposals have been made for improvements to FSBM by varying the size of blocks to more accurately match moving areas. Such methods are known as variable size block matching (VSBM) methods. Chan, Yu and Constantinides have proposed a scheme that starts with relatively large blocks, which are then repeatedly divided, this is a so-called top down approach. If the best matching error for a block is above some threshold, the block is divided into four smaller blocks, until the maximum number of blocks or locally minimum errors are obtained. The application of such top-down methods may generate block structures for an image that match real moving objects, but it seems that an approach which more directly seeks out areas of uniform motion might be more effective.

We developed a VSBM technique that detects areas of common motion, grouping them into variable sized blocks with a coding strategy based on the use of quad-trees. Use of a quad-tree obviates the need to describe the size and position of each block explicitly, only the tree description is needed. The vectors for each block in the tree are identical in nature to those of FSBM. As the process is a grouping together of smaller blocks to form larger ones, it is generally regarded as a bottom-up technique.

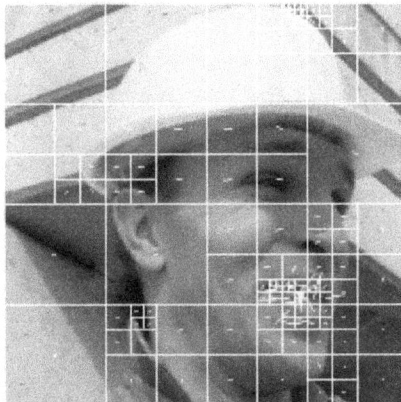

VSBM example

For the same number of blocks per frame as FSBM, our VSBM method results in a smaller mean square error (MSE), or better prediction. More significantly, for a similar level of MSE as FSBM, the VSBM technique can represent the inherent motion using fewer (variable-sized) blocks, and thus a reduced number of motion vectors.

An example of the block structure generated is shown in figure. It can be seen that the stationary background is represented by comparatively few (large) blocks, whereas the moving parts of the image are represented by smaller blocks, and hence a larger number of motion vectors.

Motion vectors are subsequently variable length coded using a quad-tree based 2d predictor mechanism.

Object Based Block-matching Motion Compensation

Evolving object-based video coding standards, such as MPEG-4, permit arbitrary-shaped objects to be encoded and decoded as separate video object planes (VOPs). There are several motivating scenarios behind the use of objects:

- where transmission bandwidth or decoder performance is limited, the user may be able to select some subset of all video objects, which are of particular interest,

- the user may wish to manipulate objects at the receiver, i.e. change position, size and depth ordering, depending on interest,

- it may be possible to replace the content of an object with material generated later or local to the receiver/display, which can be used for enhanced visualisation and "augmented reality".

VOP example

The exact method of producing VOPs from the source imagery is not defined, but it is assumed that "natural" objects are represented by shape information in addition to the usual luminance and chrominance components. Shape data is provided as a binary segmentation mask, or a grey scale alpha plane to represent multiple overlaid objects.

Figure shows a cropped frame from the MPEG-4 test sequence known as "Weather". This is made up of the synthetic weather-map, and the presenter part of the sequence for which there is an object mask or shape description.

Once the shape information has been used to derive the partial image corresponding to the object, the resulting pixel data is coded much as if it were a full image.

Fixed Size Block-Matching (FSBM)

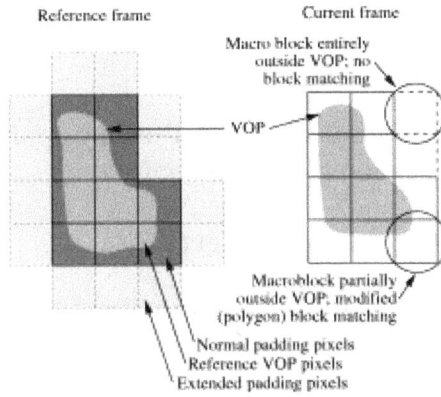

FSBM padding example

An example object is shown in figure, the object is shown in both current and reference frames, with the bounding rectangle made up of coding blocks.

The origin of the rectangle is adjusted to minimise the number of separate blocks required, although in the example this is not a particular issue. Two kinds of padding are shown:

(i) normal padding is used to fill partially populated blocks, based on a form of bi-directional interpolation,

(ii) extended padding is used to fill an extra layer of blocks by pixel replication in the appropriate direction.

Padded VOP

The padding is designed to feed the best possible input into the next stage of coding, when due to some change in the object there would be parts that do not exist in the reference frame.

Figure shows a padded example of the weather presenter object. The resulting VOP is used as the reference frame for the block matching operation, but only pixels inside the object proper are matched against this data. Motion estimation is performed by matching the 16x16 blocks within a prescribed search area of the previous frame.

For maximum accuracy, an exhaustive search is used, and matching is conducted to 1/2 pixel precision.

The block-matching process is similar to the full-frame approach described earlier. An example FSBM block structure can be seen in figure.

Variable Size Block-Matching (VSBM)

Since frame-based VSBM results in a better estimate of "true" motion and hence more efficient coding of vector information, one would expect that it can be applied to object-based systems with similar effects.

There are two problems to overcome when using a basic block matching approach to find true motion:

- The first is the majority effect, where any small area of motion inside a block will simply be lost as the matching error for the block is determined by the majority of the block. This is an argument against the use of large block sizes. Furthermore, a single block cannot effectively represent more than one motion, so there is always a trade-off between block size and error quality of match.

- The aperture problem is the second difficulty, in this case associated with small block sizes. The fewer pixels there are to match, the more spurious matches there will be due to ambiguity. Additionally there is little point in having the overhead of many small blocks if they all have the same vector (if they don't, the vectors are unlikely to all be correct).

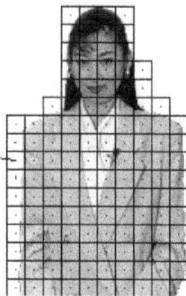

FSBM block structure   VSBM block structure

Figure shows the result of applying VSBM to the same frame of the "weather" sequence for the same quality prediction. While FSBM requires 109 blocks, VSBM only needs 44, a saving of almost 60%.

Motion vectors are then variable length coded using a differential, object-based 2D prediction strategy.

Motion compensation is an algorithmic technique used to predict a frame in a video, given the previous and/or future frames by accounting for motion of the camera and/or objects in the video. It is employed in the encoding of video data for video compression, for example in the generation of MPEG-2 files. Motion compensation describes a picture in terms of the transformation of a reference picture to the current picture. The reference picture may be previous in time or even from the future. When images can be accurately synthesised from previously transmitted/stored images, the compression efficiency can be improved.

Functionality

Motion compensation exploits the fact that, often, for many frames of a movie, the only difference between one frame and another is the result of either the camera moving or an object in the frame moving. In reference to a video file, this means much of the information that represents one frame will be the same as the information used in the next frame.

Using motion compensation, a video stream will contain some full (reference) frames; then the only information stored for the frames in between would be the information needed to transform the previous frame into the next frame.

Illustrated Example

Type	Example Frame	Description
Original		Full original frame, as shown on screen.
Difference		Differences between the original frame and the next frame.
Motion compensated difference		Differences between the original frame and the next frame, shifted right by 2 pixels. Shifting the frame *compensates* for the panning of the camera, thus there is greater overlap between the two frames.

The following is a simplistic illustrated explanation of how motion compensation works. Two successive frames were captured from the movie Elephants Dream. As can be seen from the images, the bottom (motion compensated) difference between two frames contains significantly less detail than the prior images, and thus compresses much better than the rest. Thus the information that is required to encode compensated frame will be much smaller than with the difference frame. This also means that it is also possible to encode the information using difference image at a cost of less compression efficiency but by saving coding complexity without motion compensated coding; as a matter of fact that motion compensated coding (together with motion estimation, motion compensation) occupies more than 90% of encoding complexity.

MPEG

In MPEG, images are predicted from previous frames (P frames) or bidirectionally from previous and future frames (B frames). B frames are more complex because the image sequence must be transmitted and stored out of order so that the future frame is available to generate the B frames.

After predicting frames using motion compensation, the coder finds the residual, which is then compressed and transmitted.

Global Motion Compensation

In global motion compensation, the motion model basically reflects camera motions such as:

- Dolly - moving the camera forward or backward

- Track - moving the camera left or right

- Boom - moving the camera up or down

- Pan - rotating the camera around its Y axis, moving the view left or right

- Tilt - rotating the camera around its X axis, moving the view up or down

- Roll - rotating the camera around the view axis

It works best for still scenes without moving objects.

There are several advantages of global motion compensation:

- It models the dominant motion usually found in video sequences with just a few parameters. The share in bit-rate of these parameters is negligible.

- It does not partition the frames. This avoids artifacts at partition borders.

- A straight line (in the time direction) of pixels with equal spatial positions in the frame corresponds to a continuously moving point in the real scene. Other MC schemes introduce discontinuities in the time direction.

MPEG-4 ASP supports GMC with three reference points, although some implementations can only make use of one. A single reference point only allows for translational motion which for its relatively large performance cost provides little advantage over block based motion compensation.

Moving objects within a frame are not sufficiently represented by global motion compensation. Thus, local motion estimation is also needed.

Overlapped Block Motion Compensation

Overlapped block motion compensation (OBMC) is a good solution to these problems because it not only increases prediction accuracy but also avoids blocking artifacts. When using OBMC, blocks are typically twice as big in each dimension and overlap quadrant-wise with all 8 neighbouring blocks. Thus, each pixel belongs to 4 blocks. In such a scheme, there are 4 predictions for each pixel which are summed up to a weighted mean. For this purpose, blocks are associated with a window function that has the property that the sum of 4 overlapped windows is equal to 1 everywhere.

Studies of methods for reducing the complexity of OBMC have shown that the contribution to the window function is smallest for the diagonally-adjacent block. Reducing the weight for this contribution to zero and increasing the other weights by an equal amount leads to a substantial reduction in complexity without a large penalty in quality. In such a scheme, each pixel then belongs to 3 blocks rather than 4, and rather than using 8 neighboring blocks, only 4 are used for each block to be compensated. Such a scheme is found in the H.263 Annex F Advanced Prediction mode.

Quarter Pixel (QPel) and Half Pixel Motion Compensation

In motion compensation, quarter or half samples are actually interpolated sub-samples caused by fractional motion vectors. Based on the vectors and full-samples, the sub-samples can be calculated by using bicubic or bilinear 2-D filtering.

3D Image Coding Techniques

Motion compensation is utilized in Stereoscopic Video Coding.

In video, *time* is often considered as the third dimension. Still image coding techniques can be expanded to an extra dimension.

JPEG2000 uses wavelets, and these can also be used to encode motion without gaps between blocks in an adaptive way. Fractional pixel affine transformations lead to bleeding between adjacent pixels. If no higher internal resolution is used the delta images mostly fight against the image smearing out. The delta image can also be encoded as wavelets, so that the borders of the adaptive blocks match.

2D+Delta Encoding techniques utilize H.264 and MPEG-2 compatible coding and can use motion compensation to compress between stereoscopic images.

Video Post-processing

The third stage of video development is referred to as post-production, with the first stage being pre-production, where the video is planned, and the second stage being production, where the video is filmed. Post-production generally includes everything that occurs after shooting has been completed.

Post-production services are often offered by production companies or video agencies. By definition, post-production would include editing, special effects, and/or animation. However, it could also include distribution, such as sending your video out to television corporations or uploading the video online.

There are a lot more details and knowledge that go into good post-production service than most people realize. Below is a detailed list of services that could be included in post-production service. We will discuss:

- Editing
- Colour-Correction
- Visual Effects
- Animation
- Distribution

Uses in Video Production

Video post-processing is the process of changing the perceived quality of a video on playback (done after the decoding process). Image scaling routines such as linear interpolation, bilinear interpolation, or cubic interpolation can for example be performed when increasing the size of images; this involves either subsampling (reducing or shrinking an image) or zooming (enlarging an image). This helps reduce or hide image artifacts and flaws in the original film material. It is important to understand that post-processing always involves a trade-off between speed, smoothness and sharpness.

- Image scaling and multivariate interpolation:
    - o   Nearest-neighbor interpolation

- o   linear interpolation
- o   bilinear interpolation
- o   cubic interpolation
- o   bicubic interpolation
- o   Bézier surface
- o   Lanczos resampling
- o   trilinear interpolation
- o   Tricubic interpolation
- SPP (Statistical-Post-Processing)
- Deblocking
- Deringing
- Sharpen / Unsharpen (often referred to as "soften")
- Requantization
- Luminance alterations
- Blurring / Denoising
- Deinterlacing
  - o   weave deinterlace method
  - o   bob deinterlace method
  - o   linear deinterlace method
  - o   yadif deinterlace method
- Deflicking
- 2:3 pull-down / ivtc (inverse telecine) for conversion from 24 frames/s and 23.976 frames/s to 30 frames/s and 29.97 frames/s
- 3:2 pull-up (telecine conversion) for conversion from 30 frames/s and 29.97 frames/s to 24 frames/s and 23.976 frames/s

Types of Post-Production

Editing

Editing is often the first thing that pops into people's heads when they hear 'post-production'. It is the act of selecting clips from your shoot and putting them in a specific order to tell the desired story. Editing includes both picture and sound manipulation. It would also include any lower thirds (such as a speaker's title or the title of an episode), credits, foley sound, sound effects, and soundtrack.

Colour-Correction

Colour-correction is an element of editing that is often not thought of, but it can make a huge difference to your video. When shooting a video, chances are you will be filming with multiple cameras and in multiple locations. It is production's job to make sure lighting and camera settings are as consistent as possible when filming, but there will always be a variation in lighting and colour depth in post. Colour-correction helps to maintain continuity in your story and sets the tone in your video.

Visual Effects

Visual effects is another category that can include several different things. Visual effects are visuals done in post-production such as green-screen compositing or muzzle flashes. The term can refer to any image that could not be filmed during production for whatever reason.

CGI, or computer generated images, is a type of visual effect where characters, models, or designs are created using a computer. It is the most common form of visual effects in commercial filmmaking, and takes talent and time.

Keep in mind, visual effects are not to be confused with special effects, which are effects that are implemented in production such as prosthetic makeup or on-set explosions.

Animation

Animation is another post-production service an agency may offer. By definition, animation is the process of making the illusion of motion by using rapid succession of sequential images that differ from each otheronly minimally. Most animation is done via computer nowadays, technically making it CGI, but there is an unspoken distinction between animation and CGI. Usually CGI refers to images that are meant to look life-like, while animation is not masquerading as 'real-life'.

Distribution

Video distribution is another category that can refer to several things. Platforms can range from streaming services such as Netflix and Amazon, to social medias like Facebook and Instagram, to television channels like Comedy Central or MTV.

Social media platforms each have their own video requirements, such as aspect ration and maximum length. This can impact how many versions of a video would need to be exported after editing and colour correction.

Streaming services use a non-linear (or unicast) distribution model, which includes time-shifted viewing (like Sky Catch-up TV) and on-demand viewing (such as You-Tube).

Then you have television distribution, which is an entirely different process entirely. TV uses the linear broadcast model, where the content is distributed centrally from the broadcaster at a scheduled time. Not to mention, your video would need to pass Clearcast and Adstream inspection before they can be aired.

For each of these broadcasting models (just like social media), there are different technical specifications your video must meet. Post-production distribution services could include taking care of these requirements and inspections for you.

In conclusion, post-production services can include a myriad of things, ranging from editing, which consists of the basic arrangement of visual and audio material, colour-correction to add the finishing touches, visual effects to address any creative needs, animation, and distribution, to help your video be seen. Hiring a knowledgeable media company can make the post-production process infinitely easier for your business. Not only can it save you time, but it will also ensure the correct steps are taken to produce the best quality edit possible, and ensure your video is seen by as many people as possible.

References

- Wang, Yao, Jörn Ostermann, and Ya-Qin Zhang. Video Processing and Communications. Signal Processing Series. Upper Saddle River, N.J.: Prentice Hall, 2002. ISBN 0-13-017547-1

- US patent US4220969 / CA1103345A1, Kazuhiko Nitadori, "Digital scan converter", published September 2, 1980. assigned to OKI Electric Ind. C. Ltd. Retrieved August 19, 2013

- Chen, Yangkang (2017). "Probing the subsurface karst features using time-frequency decomposition". Interpretation. 4 (4): T533-T542. doi:10.1190/INT-2016-0030.1

- Chen, Yangkang (2017). "Fast dictionary learning for noise attenuation of multidimensional seismic data". Geophysical Journal International. 209 (1): 21–31. Bibcode:2017GeoJI.209...21C. doi:10.1093/gji/ggw492

- Jung, J.H.; Hong, S.H. (2011). "Deinterlacing method based on edge direction refinement using weighted maximum frequent filter". Proceedings of the 5th International Conference on Ubiquitous Information Management and Communication. ACM. ISBN 978-1-4503-0571-6

- "EBU R115-2005: FUTURE HIGH DEFINITION TELEVISION SYSTEMS" (PDF). EBU. May 2005. Archived (PDF) from the original on 2009-05-27. Retrieved 2009-05-24

- Chen, Yangkang; Fomel, Sergey (November–December 2015). "Random noise attenuation using local signal-and-noise orthogonalization". Geophysics. 80 (6): WD1-WD9. doi:10.1190/GEO2014-0227.1

- Burwen, Richard S. (December 1971). "Design of a Noise Eliminator System". Journal of The Audio Engineering Society. 19: 906–911

- Kallenberger, Richard H., Cvjetnicanin, George D. (1994). Film into Video: A Guide to Merging the Technologies. Focal Press. ISBN 0-240-80215-2

- Belton, John (Spring 2008). "Painting by the Numbers: The Digital Intermediate". Film Quarterly. 61 (3): 58–65. doi:10.1525/fq.2008.61.3.58

- Besag, Julian (1986). "On the Statistical Analysis of Dirty Pictures". Journal of the Royal Statistical Society. Series B (Methodological). 48 (3): 259–302. JSTOR 2345426

5

Video Editing: Types and Techniques

Video editing refers to the arrangement and manipulation of video shots. It is significant in films and television shows, video essays and video advertisements. The various technological innovations in video editing, such as continuity editing, linear and non-linear editing systems have been covered in great detail in this chapter.

Video editing is the process of manipulating video by rearranging different shots and scenes in order to create a whole new output. It can be as simple as stitching together different scenes and shots with simple video transitions, and can become as complicated as adding different computer-generated imagery (CGI), audio and tying together different elements, which may take years, thousands of man-hours and millions of dollars to accomplish, as is the case with big-budget motion pictures.

Video editing is the process of putting together, cleaning up and finalizing a video for presentation or output. It is mostly used to describe post-production work, which is the work done after all of the shots and footage have been taken and all that still needs to be done is put them together in order to come up with the final output.

Video editing, however, is mostly used to refer to amateur productions and other professional yet small-scale works such as in TV stations and news networks. In contrast, for professional cinema and Hollywood productions, video editing is only a small part of post-production work.

We use the word *editing* to mean any of the following:

- Rearranging, adding and/or removing sections of video clips and/or audio clips.
- Applying colour correction, filters and other enhancements.
- Creating transitions between clips.

Types of Editing

Once the province of expensive machines called video editors, video editing software is now available for personal computers and workstations. Video editing includes cutting segments (trimming), re-sequencing clips, and adding transitions and other Special Effects.

- Linear video editing, using video tape and is edited in a very linear way. Several video clips from different tapes are recorded to one single tape in the order that they will appear.

- Non-linear editing system (NLE), This is edited on computers with specialised software. These are non destructive to the video being edited and use programs such as Adobe Premiere Pro, Final Cut Pro and Avid.

- Offline editing is the process in which raw footage is copied from an original source, without affecting the original film stock or video tape. Once the editing has been completely edited, the original media is then re-assembled in the online editing stage.

- Online editing is the process of reassembling the edit to full resolution video after an offline edit has been performed and is done in the final stage of a video production.

- Vision mixing, when working within live television and video production environments. A vision mixer is used to cut live feed coming from several cameras in real time.

Background

Video editing is the process of editing segments of motion video production footage, special effects and sound recordings in the post-production process. Motion picture film editing is a predecessor to video editing and, in several ways, video editing simulates motion picture film editing, in theory and the use of linear video editing and video editing software on non-linear editing systems (NLE). Using video, a director can communicate non-fictional and fictional events. The goals of editing is to manipulate these events to bring the communication closer to the original goal or target. It is a visual art.

Early 1950s video tape recorders (VTR) were so expensive, and the quality degradation caused by copying was so great, that 2 inch Quadruplex videotape was edited by visualizing the recorded track with ferrofluid and cutting with a razor blade or guillotine cutter and splicing with video tape. The two pieces of tape to be joined were painted with a solution of extremely fine iron filings suspended in carbon tetrachloride, a toxic and carcinogenic compound. This "developed" the magnetic tracks, making them visible when viewed through a microscope so that they could be aligned in a splicer designed for this task.

Improvements in quality and economy, and the invention of the flying erase-head, allowed new video and audio material to be recorded over the material already recorded on an existing magnetic tape and was introduced into the linear editing technique. If a scene closer to the beginning of the video tape needed to be changed in length, all later scenes would need to be recorded onto the video tape again in sequence. In addition, sources could be played back simultaneously through a vision mixer (video switcher) to create more complex transitions between scenes. A popular 1970-80s system for doing that was the U-matic equipment (named for the U-shaped tape path). That system used two tape players and one tape recorder, and edits were done by automatically having the machines back up, then speed up together in synchrony, so the edit didn't roll or glitch. Later, 1980-90's came the smaller beta equipment (named for the B-shaped tape path), and more complex controllers, some of which did the synchronizing electronically.

Editor in linear VCR suite

There was a transitional analog period using multiple source videocassette recorder (VCR)s with the EditDroid using LaserDisc players, but modern NLE systems edit video digitally captured onto a hard drive from an analog video or digital video source. Content is ingested and recorded natively with the appropriate codec that the video editing software uses to process captured footage. High-definition video is becoming more popular and can be readily edited using the same video editing software along with related motion graphics programs. Video clips are arranged on a timeline, music tracks, titles, digital on-screen graphics are added, special effects can be created, and the finished program is "rendered" into a finished video. The video may then be distributed in a variety of ways including DVD, web streaming, QuickTime Movies, iPod, CD-ROM, or video tape.

Home Video Editing

Like some other technologies, the cost of video editing has declined by an order of magnitude or more. The original 2" Quadruplex system cost so much that many television production facilities could only afford a single unit and editing was a highly involved process requiring special training. In contrast to this, nearly any home computer sold since the year 2000 has the speed and storage capacity to digitize and edit standard-definition television (SDTV). The two major retail operating systems include basic video editing software - Apple's iMovie and Microsoft's Windows Movie Maker. Additional options exist, usually as more advanced commercial products. As well as these commercial products, there are opensource video-editing programs. Automatic video editing products have also emerged, opening up video editing to a broader audience of amateurs and reducing the time it takes to edit videos. These exist usually as media storage services, such as Google with its Google Photos or smaller companies like Vidify.

Goals of Editing

There are many reasons to edit a video and your editing approach will depend on the desired outcome. Before you begin you must clearly define your editing goals, which could include any of the following:

- Remove Unwanted Footage

This is the simplest and most common task in editing. Many videos can be dramatically improved by simply getting rid of the flawed or unwanted bits.

- Choose the Best Footage

It is common to shoot far more footage than you actually need and choose only the best material for the final edit. Often you will shoot several versions (takes) of a shot and choose the best one when editing.

- Create a Flow

Most videos serve a purpose such as telling a story or providing information. Editing is a crucial step in making sure the video flows in a way which achieves this goal.

- Add Effects, Graphics, Music, etc

This is often the "wow" part of editing. You can improve most videos (and have a lot of fun) by adding extra elements.

- Alter the Style, Pace or Mood of the Video

A good editor will be able to create subtle mood prompts in a video. Techniques such as mood music and visual effects can influence how the audience will react.

- Give the Video a Particular "Angle"

Video can be tailored to support a particular viewpoint, impart a message or serve an agenda.

Continuity Editing

Filmmakers and editors may work with various goals in mind. Traditionally, commercial cinema prefers the continuity system, or the creation of a logical, continuous narrative which allows the viewer to suspend disbelief easily and comfortably. Continuity Editing is the process of creating a smooth and seamless narrative experience for the audience- it can be useful to think of it as invisible editing.

Continuity editing is the predominant style of film editing and video editing in the post-production process of filmmaking of narrative films and television programs. The purpose of continuity editing is to smooth over the inherent discontinuity of the editing process and to establish a logical coherence between shots.

Common Techniques of Continuity Editing

Continuity editing can be divided into two categories: temporal continuity and spatial continuity. Within each category, specific techniques will work against a sense of continuity. In other words, techniques can cause a passage to be continuous, giving the viewer a concrete physical narration to follow, or discontinuous, causing viewer disorientation, pondering, or even subliminal interpretation or reaction, as in the montage style.

The important ways to preserve temporal continuity are avoiding the ellipsis, using continuous diegetic sound, and utilizing the match on action technique.

An ellipsis is an apparent break in natural time continuity as it is implied in the film's story. The simplest way to maintain temporal continuity is to shoot and use all action involved in the story's supposed duration whether it be pertinent or not. It would also be necessary to shoot the whole film in one take in order to keep from having to edit together different shots, causing the viewer's temporal disorientation. However, in a story which is to occupy many hours, days, or years, a viewer would have to spend too long watching the film. So although in many cases the ellipsis would prove necessary, elimination of it altogether would best preserve any film's temporal continuity.

Diegetic sound is that which is to have actually occurred within the story during the action being viewed. It is sound that comes from within the narrative world of a film (including off-screen sound). Continuous diegetic sound helps to smooth temporally questionable cuts by overlapping the shots. Here the logic is that if a sonic occurrence within the action of the scene has no breaks in time, then it would be impossible for the scene and its corresponding visuals to be anything but temporally continuous.

Match on action technique can preserve temporal continuity where there is a uniform, unrepeated physical motion or change within a passage. A match on action is when some action occurring before the temporally questionable cut is picked up where the cut left it by the shot immediately following. For example, a shot of someone tossing a ball can be edited to show two different views, while maintaining temporal continuity by being sure that the second shot shows the arm of the subject in the same stage of its motion as it was left when cutting from the first shot.

Temporal discontinuity can be expressed by the deliberate use of ellipses. Cutting techniques useful in showing the nature of the specific ellipses are the dissolve and the fade. Other editing styles can show a reversal of time or even an abandonment of it altogether. These are the flashback and the montage techniques, respectively.

A fade is a gradual transformation of an image to or back from black. A dissolve is a simultaneous overlapping transition from one shot to another that does not involve an instantaneous cut or change in brightness. Both forms of transition (fade and dissolve) create an ambiguous measure of ellipsis that may constitute diagetic (narrative) days, months, years or even centuries. Through the use of the dissolve or the fade, one may allude to the relative duration of ellipses where the dissolve sustains a visual link but the fade to black does not. It cannot be argued that one constitutes short ellipsis and the other long however, as this negates the very functional ambiguity created by such transitions. Ambiguity is removed through the use of captions and intertitles such as "three weeks later" if desired.

The flashback is a relocation of time within a story, or more accurately, a window through which the viewer can see what happened at a time prior to that considered (or assumed) to be the story present. A flashback makes its time-frame evident through the scene's action or through the use of common archetypes such as sepia toning, the use of home-movie style footage, period costume or even through obvious devices such as clocks and calendars or direct character linkage. For example, if after viewing a grown man in the story present, a cut to a young boy being addressed by the man's name occurs, the viewer can assume that the young boy scene depicts a time previous to the story present. The young boy scene would be a flashback.

The montage technique is one that implies no real temporal continuity whatsoever. Montage is achieved with a collection of symbolically related images, cut together in a way that suggests psychological relationships rather a temporal continuum.

Just as important as temporal continuity to overall continuity of a film is spatial continuity. And like temporal continuity, it can be achieved a number of ways: the establishing shot, the 180 degree rule, the eyeline match, and match on action.

The establishing shot is one that provides a view of all the space in which the action is occurring. Its theory is that it is difficult for a viewer to become disoriented when all the story space is presented before him. The establishing shot can be used at any time as a reestablishing shot. This might be necessary when a complex sequence of cuts may have served to disorient the viewer.

One way of preventing viewer disorientation in editing is to adhere to the 180 degree rule. The rule prevents the camera from crossing the imaginary line connecting the subjects of the shot. Another method is the eyeline match. When shooting a human subject, he or she can look towards the next subject to be cut to, thereby using the former's self as a reference for the viewer to use while locating the new subject within the set.

With the establishing shot, 180 degree rule, eyeline match, and the previously discussed match on action, spatial continuity is attainable; however, if wishing to convey a disjointed space, or spatial discontinuity, aside from purposefully contradicting the continuity tools, one can take advantage of crosscutting and the jump cut.

Cross-cutting is a technique which conveys an undeniable spatial discontinuity. It can be achieved by cutting back and forth between shots of spatially unrelated places. In these cases, the viewer will understand clearly that the places are supposed to be separate and parallel. So in that sense, the viewer may not become particularly disoriented, but under the principle of spatial continuity editing, crosscutting is considered a technique of spatial discontinuity.

The jump cut is undoubtedly a device of disorientation. The jump cut is a cut between two shots that are so similar that a noticeable jump in the image occurs. The 30 degree rule was formulated for the purpose of eliminating jump cuts. The 30 degree rule requires that no edit should join two shots whose camera viewpoints are less than 30 degrees from one another.

The 180 Degree Rule and Spatial Continuity

The first rule that any filmmaker needs to learn before he picks up his camera is the 180 degree rule. Adherence to this rule is necessary to maintain continuity in your scene. What you do is create an imaginary line across your set that you will not cross with the camera. This way if the actor is on the left side of the frame and the actress is one right side in the master shot, they will stay in those established positions throughout the scene as the medium shots and close ups are editing together.

If the camera crossed the line and the actress appeared frame left and the actor frame right, then this would cause the audience to become disoriented because the established spatial continuity had been violated. Once the spatial distance and positions has been established, you should not violate it if you want to maintain continuity.

Cutting on the Action and Temporal Continuity

Entire scenes and montages can move between time, but the shots that compose the scene should have temporal continuity. An individual scene needs to feel as if it is happening right now in real time. The most common way of maintaining this illusion is to cut your shots on actions so that they match up to each other.

For example, let's say that we're editing two shots together of a man throwing a football. We can start with the close up where he begins to throw the ball and then cut to the wide shot where we see the ball leaving his hand and traveling across the field. We would want to cut the two shots together so that they meet at a point when the man's arm is in the same position. This way the action appears to be seamless when edited together.

Breaking the Rules

Not everything in your movie needs to be edited to the continuity style. Montages are a common example of a sequence of shots not edited for continuity. Instead they are put together to show a great passage of time while having a psychological effect on the audience by progressing the story.

Even some of the most celebrated filmmakers break the rules. In Jean Luc Goddard's 'Breathless', the director uses jump cuts to show the passage of time. But, if you plan on making a career in filmmaking, you should first learn the rules before you decide to break them.

Linear Video Editing

Linear video editing describes a process in which scenes are copied from one video tape to another, using two tape VCRs, in the order required. The new tape is thus created in a linear fashion. The disadvantage of this method is that it is not possible to insert or delete scenes from the new tape without re-copying all the subsequent scenes. Linear editing was the method originally used with analogue video tapes.

It is a process of selecting, arranging and modifying images and sound in a pre-determined, ordered sequence – from start to finish. Linear editing is most commonly used when working with videotape. Unlike film, videotape cannot be physically cut into pieces to be spliced together to create a new order. Instead, the editor must dub or record each desired video clip onto a master tape.

For example, let's say an editor has three source tapes; A, B and C and he decided that he would use tape C first, B second and A third. He would then start by cutting up tape C to the beginning of the clip he wants to use, then as he plays tape C, it would also be simultaneously recording the clip onto a master tape. When the desired clip from tape C is done, the recording is stopped. Then the whole process is repeated with tapes B and A.

Early Technology

The first widely accepted video tape in the United States was two-inch quadruplex videotape and travelled at 15 inches per second. To gain enough head-to-tape speed, four video recording and playback heads were spun on a head wheel across most of the two-inch width of the tape. (Audio and synchronization tracks were recorded along the sides of the tape with stationary heads.) This system was known as "quad" (for "quadruplex") recording.

The resulting video tracks were slightly less than a ninety-degree angle (considering the vector addition of high-speed spinning heads tracing across the 15 inches per second forward motion of the tape).

Originally, video was edited by visualizing the recorded track with ferrofluid and cutting it with a razor blade or guillotine cutter and splicing with video tape, in a manner similar to film editing. This was an arduous process and avoided where possible. When it was used, the two pieces of tape to be joined were painted with a solution of extremely fine iron filings suspended in carbon tetrachloride, a toxic and carcinogenic compound. This "developed" the magnetic tracks, making them visible when viewed through a microscope so that they could be aligned in a splicer designed for this task. The tracks had to be cut during a vertical retrace, without disturbing the odd-field/even-field ordering. The cut also had to be at the same angle that the video tracks were laid down on the tape. Since the video and audio read heads were several inches apart it was not possible to make a physical edit that would function correctly in both video and audio. The cut was made for video and a portion of audio then re-copied into the correct relationship, the same technique as for editing 16mm film with a combined magnetic audio track.

The disadvantages of physically editing tapes were many. Some broadcasters decreed that edited tapes could not be reused, in an era when the relatively high cost of the machines and tapes was balanced by the savings involved in being able to wipe and reuse the media. Others, such as the BBC, allowed reuse of spliced tape in certain circumstances as long as it conformed to strict criteria about the number of splices in a given duration, usually a maximum of five splices for every half hour. The process required great skill, and often resulted in edits that would roll (lose sync) and each edit required several minutes to perform, although this was also initially true of the electronic editing that came later.

In the USA, the 1961-62 Ernie Kovacs ABC specials and *Rowan & Martin's Laugh-In* were the only TV shows to make extensive use of splice editing of videotape.

Introduction of Computerized Systems

A system for editing Quad tape "by hand" was developed by the 1960s. It was really just a means of synchronizing the playback of two machines so that the signal of the new shot could be "punched in" with a reasonable chance at success. One problem with this and early computer-controlled systems was that the audio track was prone to suffer artifacts (i.e. a short buzzing sound) because the video of the newly recorded shot would record into the side of the audio track. A commercial solution known as "Buzz Off" was used to minimize this effect.

For more than a decade, computer-controlled Quad editing systems were the standard post-production tool for television. Quad tape involved expensive hardware, time-consuming setup, relatively long rollback times for each edit and showed misalignment as disagreeable "banding" in the video. However, it should be mentioned that Quad tape has a better bandwidth than any smaller-format analogue tape, and properly handled could produce a picture indistinguishable from that of a live camera.

Further Advancement in Technology

A Sony BVE-910 linear editing system's keyboard

When helical scan video recorders became the standard it was no longer possible to physically cut the tape. At this point video editing became a process of using two video tape machines, playing back the source tape (or "raw footage") from one machine and copying just the portions desired on to a second tape (the "edit master").

The bulk of linear editing is done simply, with two machines and an edit controller device to control them. Many video tape machines are capable of controlling a second machine, eliminating the need for an external editing control device.

This process is "linear", rather than non-linear editing, as the nature of the tape-to-tape copying requires that all shots be laid out in the final edited order. Once a shot is on tape, nothing can be placed ahead of it without overwriting whatever is there already. If absolutely necessary, material can be dubbed by copying the edited content onto another tape, however as each copy generation degrades the image cumulatively, this is not desirable.

One drawback of early video editing technique was that it was impractical to produce a rough cut for presentation to an Executive producer. Since Executive Producers are never familiar enough with the material to be able to visualise the finished product from inspection of an edit decision list (EDL), they were deprived of the opportunity to voice their opinions at a time when those opinions could be easily acted upon. Thus, particularly in documentary television, video was resisted for quite a long time.

Peak usage

Video editing reached its full potential in the late 1970s when computer-controlled minicomputer edit controllers along with communications protocols were developed, which could orchestrate an edit based on an EDL, using timecode to synchronize multiple tape machines and auxiliary devices using a 9-Pin Protocol. The most popular and widely used computer edit systems came from Sony, Ampex and the venerable CMX. Systems such as these were expensive, especially when considering auxiliary equipment like VTR, video switchers and character generators (CG) and were usually limited to high-end post-production facilities.

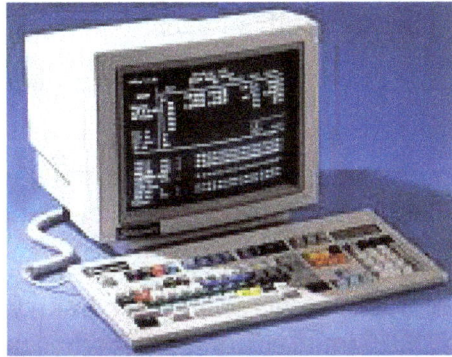

Strassner Editing Systems

Jack Calaway of Calaway Engineering was the first to produce a lower-cost, PC-based, "CMX-style" linear editing system which greatly expanded the use of linear editing systems throughout the post-production industry. Following suit, other companies, including EMC and Strassner Editing Systems, came out with equally useful competing editing products.

Current Usage

While computer based video editing software has been adopted throughout most of the commercial, film, industrial and consumer video industries, linear video tape editing is still commonplace in television station newsrooms for the production of television news, and medium-sized production facilities which haven't made the capital investment in newer technologies. News departments often still use linear editing because they can start editing tape and feeds from the field as soon as received since no additional time is spent capturing material as is necessary in non-linear editing systems and systems that are able to digitally record and edit simultaneously have only recently become affordable for small operations.

Pros vs Cons

There are a couple of disadvantages one would come across when using the linear video editing method. First, it is not possible to insert or delete scenes from the master tape without re-copying all the subsequent scenes. As each piece of video clip must be laid down in real time, you would not be able to go back to make a change without re-editing everything after the change.

Secondly, because of the overdubbing that has to take place if you want to replace a current clip with a new one, the two clips must be of the exact same length. If the new clip is too short, the tail end of the old clip will still appear on the master tape. If it's too long, then it'll roll into the next scene. The solution is to either make the new clip fit to the current one, or rebuild the project from the edit to the end, both of which is not very pleasant. Meanwhile, all that overdubbing also causes the image quality to degrade.

However, linear editing still has some advantages:

- It is simple and inexpensive. There are very few complications with formats, hardware conflicts, etc.

- For some jobs linear editing is better. For example, if all you want to do is add two sections of video together, it is a lot quicker and easier to edit tape-to-tape than to capture and edit on a hard drive.

- Learning linear editing skills increases your knowledge base and versatility. According to many professional editors, those who learn linear editing first tend to become better all-round editors.

Online Editing

Online editing is the processing of video or graphic editing which is done as the final step of video making. This step is the opposite of offline editing, where a video is processed in its earliest and most raw state. In small productions, online-offline workflow is replaced with video editing software working on a non-linear editing system (NLE), whereas sophisticated post-productions that utilize high-grade equipment still make use of offline-online workflow.

The term online originated in the telecommunication industry, meaning "Under the direct control of another device" (automation). Online editors such as the Sony BVE-9000 edit control unit used the RS-422 remote control 9-Pin Protocol to allow the computer-interface of edit controllers to control video tape recorders (VTR) via a series of commands. The protocol supports a variety of devices including one-inch reel-to-reel type C videotape as well as videocassette recorders (VCR) to Fast-Forward, Rewind and Play and Record based on SMPTE timecode. The controllers have the ability to interface with professional audio equipment like audio mixers with console automation.

The video quality first introduced with Avid's Media Composer in 1989 was incapable of producing broadcast quality images due to computer processing limitations. The term 'Online' changed from its original meaning to where the pictures are re-assembled at full or 'online' resolution. An edit decision list (EDL) or equivalent is used to carry over the cuts and dissolves created during the offline edit. This conform is checked against a video copy of the offline edit to verify that the edits are correct and frame-accurate. This workprint (cutting copy in the UK) also provides a reference for any digital video effects that need to be added.

After conforming the project, the online editor will add visual effects, lower third titles, and apply color correction. This process is typically supervised by the client(s). The editor will also ensure that the program meets the technical delivery broadcast safe specs of the broadcaster, ensuring proper video levels, aspect ratio, and blanking width.

Sometimes the online editor will package the show, putting together each version. Each version may have different requirements for the formatting (i.e. closed blacks), bumper music, use of a commercial bumper, different closing credits, etc.

Projects may be re-captured at the lowest level of video compression possible - ideally with no compression at all.

Online editing is the final cut of an edited video. The difference in video editing techniques was distinguished in the days of analog video processing. This was because physical tape would wear out from constant running and rolling back and forth; pushing the editors to divide the task in high- and low-profile processes. With the introduction of digital media, this difference begin to diminish because the video quality is unaffected by constant playing, and software is powerful enough to handle offline and online editing processes. However, very high quality master tapes for exclusive productions still use online editing on their videos.

Non-linear Editing System

Non-linear video editing is achieved by loading the video material into a computer from analogue or digital tape. The editing process creates a new 'tape' by storing all the commands entered by the operator. This method allows the operator to cut, copy and paste scenes in any order and make any changes desired. At the completion of the editing process the computer can then build a new file by applying the commands to the original digital image stored on the disk. The original digital image on the disk is unchanged. The new video file can then be outputted to a video tape, attached to an email or posted to the web.

It is a way of random access editing, which means instant access to whatever clip you want, whenever you want it. So instead of going in a set order, you are able to work on any segment of the project at any time, in any order you want. In nonlinear video editing, the original source files are not lost or modified during editing. This is done through an edit decision list (EDL), which records the decisions of the editor and can also be interchanged with other editing tools. As such, many variations of the original source files can exit without needing to store many different copies, allowing for very flexible editing. It is also easy to change cuts and undo previous decisions simply by editing the EDL, without having to have the actual film data duplicated. Loss of video quality is also avoided due to not having to repeatedly re-encode the data when different effects are applied.

Basic Techniques

A non-linear editing approach may be used when all assets are available as files on video servers or hard disks, rather than recordings on reels or tapes—while linear editing is tied to the need to sequentially view film or hear tape. Non-linear editing enables direct access to any video frame in a digital video clip, without needing to play or scrub/shuttle through adjacent footage to reach it, as is necessary with video tape linear editing systems.

When ingesting audio or video feeds, metadata are attached to the clip. Those metadata can be attached automatically (timecode, localization, take number, name of the clip) or manually (players names, characters, in sports: red card, goal). It is then possible to access any frame by entering directly the timecode or the descriptive metadata. An editor can, for example at the end of the day in the Olympic Games, easily retrieve all the clips related to the players who received a gold medal.

The non-linear editing method is similar in concept to the cut and paste techniques used in IT. However, with the use of non-linear editing systems, the destructive act of cutting of film negatives is eliminated. It can also be viewed as the audio/video equivalent of word processing, which is why it is called desktop video editing in the consumer space.

Broadcast Workflows and Advantages

Video and audio data are first captured to hard disk-based systems, or other digital storage devices. The data is then imported into servers employing any necessary transcoding, digitizing or transfer. Once imported, the source material can be edited on a computer using application software, any of a wide range of video editing software.

Editing software records the editor's decisions in an edit decision list (EDL) that is exportable to other editing tools. Many generations and variations of the original source files can exist without storing many different copies, allowing for very flexible editing. It also makes it easy to change cuts and undo previous decisions simply by editing the edit decision list (without having to have the actual film data duplicated). Generation loss is also controlled, due to not having to repeatedly re-encode the data when different effects are applied.

Compared to the linear method of tape-to-tape editing, non-linear editing offers the flexibility of film editing, with random access and easy project organization. In non-linear editing, the original source files are not lost or modified during editing. This is one of the biggest advantages of non-linear editing compared to linear editing. With the edit decision lists, the editor can work on low-resolution copies of the video. This makes it possible to edit both standard-definition broadcast quality and high definition broadcast quality very quickly on desktop computers that may not have the power to process huge full-quality high-resolution data in real-time.

The costs of editing systems have dropped such that non-linear editing tools are now within the reach of home users. Some editing software can now be accessed free as web applications; some, like Cinelerra (focused on the professional market) and Blender3D, can be downloaded as free software; and some, like Microsoft's Windows Movie Maker or Apple Inc.'s iMovie, come included with the appropriate operating system.

Accessing the Material

The non-linear editing retrieves video media for editing. Because these media exist on the video server or other mass storage that stores the video feeds in a given codec, the editing system can use several methods to access the material:

Direct access

> The video server records feeds with a codec readable by the editing system, has network connection to the editor and allows direct editing. The editor previews material directly on the server (which it sees as remote storage) and edits directly on the server without transcoding or transfer.

Shared storage

> The video server transfers feeds to and from shared storage that is accessible by all editors. Media in the appropriate codec on the server need only transferred. If recorded with a different codec, media must be transcoded during transfer. In some cases (depending on material), files on shared storage can be edited even before the transfer is finished.

Importing

> The editor downloads the material and edits it locally. This method can be used with the previous methods.

Repartition of Editors Brands in Broadcast Industry

According to reports, *Avid's Media Composer is still the most-used NLE on prime-time TV productions, being employed on up to 90 percent of evening broadcast shows. Apple's Final Cut Pro 7 software is being used by 54.6 percent of the professional editing community.* Globally, Avid Media Composer was once more used by broadcasters, and

Final Cut Pro more by domestic users. This positioning has changed, and many more editing platforms now exist.

Domestic usage

A multimedia computer for non-linear editing of video may have a video capture card to capture analog video and/or a FireWire connection to capture digital video from a DV camera, with its video editing software. Modern web-based editing systems can take video directly from a camera phone over a GPRS or 3G mobile connection, and editing can take place through a web browser interface, so, strictly speaking, a computer for video editing does not require any installed hardware or software beyond a web browser and an internet connection.

Various editing tasks can then be performed on the imported video before it is exported to another medium, or MPEG encoded for transfer to a DVD.

Quality

At one time, a primary concern with non-linear editing had been picture and sound quality. Storage limitations at the time required that all material undergo lossy compression techniques to reduce the amount of memory occupied.

Improvements in compression techniques and disk storage capacity have mitigated these concerns, and the migration to High Definition video and audio has virtually removed this concern completely. Most professional NLEs are also able to edit uncompressed video with the appropriate hardware.

Pros vs Cons

There are many advantages a nonlinear video editing system presents. First, it allows you access to any frame, scene, or even groups of scenes at any time. Also, as the original video footage is kept intact when editing, you are able to return to the original take whenever you like. Secondly, nonlinear video editing systems offers the flexibility of editing. You can change your mind a hundred times over and changes can also be made a hundred times over without having to start all over again with each change. Thirdly, it is also possible to edit both standard definition (SD) and high definition (HD) broadcast quality videos very quickly on normal PCs which do not have the power to do the full processing of the huge full quality high resolution data in real-time.

The biggest downside to nonlinear video editing is the cost. While the dedicated hardware and software doesn't cost much, the computers and hard drives do, from two to five times more than the gear. As such, the average price for a basic nonlinear video editing package can come in between $5,000 and $10,000. For stand-alone systems that approach broadcast quality, the amount you pay may be twice that. However, as the nonlinear technology pushes forward, count on big gains in digital video storage and compression, as well as lower prices on computers and hard disks in the very near future.

Offline Editing

Technology continues to grow and push the boundaries of what's possible in the world of video editing. Screen resolution continues to increase, effects become more verbose, and the upper limit of editing computers is stretched to the max. Video editors need to be prepared for the edge of the envelope and offline editing can help them swallow the biggest of projects with ease.

Offline editing in simple terms is the use of proxy footage, duplicate footage of the original source, for video editing. The original video files are not used in the editing process, instead footage that is lower in resolution, with a smaller file size, and thus a lesser data rate, is used. After the offline edit is completed using the proxy footage, the online edit is conformed to the offline edit. The online edit is created by taking the timecode from the offline edit and applying it to the corresponding source footage, making a duplicate edit at full resolution. The offline edit also shares transitions and effects with the online edit for a complete re-creation of a video editor's work.

Modern offline video editing is conducted in a non-linear editing (NLE) suite. The digital revolution has made the offline editing workflow process immeasurably quicker, as practitioners moved from time-consuming (video tape to tape) linear video editing online editing suites, to computer hardware and video editing software such as Adobe Premiere, Final Cut Pro, Avid, Sony Vegas, Lightworks and VideoPad. Typically, all the original footage (often tens or hundreds of hours) is digitized into the suite at a low resolution. The editor and director are then free to work with all the options to create the final cut.

The goal of offline editing is to create a EDL (edit decision list) that will be used in putting together the final online version of a production. This can be done with relatively inexpensive equipment using low-resolution copies of the original footage.

In the offline phase a rough cut can be shown to a director, producer, or sponsor for approval. Typically, at this point a number of changes will be made.

An important part of the creative process is trying out many creative possibilities. Hours can be spent on just a few minutes, or even a few seconds, of a production. This can become prohibitive expensive if full online facilities and personnel are used.

Once the major decisions are made offline, the EDL that is generated can be taken to editing personnel skilled in color balancing scenes, audio sweetening, smoothing out transitions, visual effects, etc., to put together the final (online) version of the production.

However, when time is limited and optimum technical and artistic quality are not major concerns, you can skip over the offline phase.

For example, in preparing a news segment for broadcast even a high-quality laptop computer equipped with one of the many available editing programs can be used to create a final news package for broadcast.

Permissions

All chapters in this book are published with permission under the Creative Commons Attribution Share Alike License or equivalent. Every chapter published in this book has been scrutinized by our experts. Their significance has been extensively debated. The topics covered herein carry significant information for a comprehensive understanding. They may even be implemented as practical applications or may be referred to as a beginning point for further studies.

We would like to thank the editorial team for lending their expertise to make the book truly unique. They have played a crucial role in the development of this book. Without their invaluable contributions this book wouldn't have been possible. They have made vital efforts to compile up to date information on the varied aspects of this subject to make this book a valuable addition to the collection of many professionals and students.

This book was conceptualized with the vision of imparting up-to-date and integrated information in this field. To ensure the same, a matchless editorial board was set up. Every individual on the board went through rigorous rounds of assessment to prove their worth. After which they invested a large part of their time researching and compiling the most relevant data for our readers.

The editorial board has been involved in producing this book since its inception. They have spent rigorous hours researching and exploring the diverse topics which have resulted in the successful publishing of this book. They have passed on their knowledge of decades through this book. To expedite this challenging task, the publisher supported the team at every step. A small team of assistant editors was also appointed to further simplify the editing procedure and attain best results for the readers.

Apart from the editorial board, the designing team has also invested a significant amount of their time in understanding the subject and creating the most relevant covers. They scrutinized every image to scout for the most suitable representation of the subject and create an appropriate cover for the book.

The publishing team has been an ardent support to the editorial, designing and production team. Their endless efforts to recruit the best for this project, has resulted in the accomplishment of this book. They are a veteran in the field of academics and their pool of knowledge is as vast as their experience in printing. Their expertise and guidance has proved useful at every step. Their uncompromising quality standards have made this book an exceptional effort. Their encouragement from time to time has been an inspiration for everyone.

The publisher and the editorial board hope that this book will prove to be a valuable piece of knowledge for students, practitioners and scholars across the globe.

Index

www.ingramcontent.com/pod-product-compliance
Lightning Source LLC
Chambersburg PA
CBHW061948190326
41458CB00009B/2819